UNIPA Springer Series

The **UNIPA Springer Series** publishes single and co-authored thematic collected volumes, monographs, handbooks and advanced textbooks on specific issues of particular relevance in six core scientific areas. The issues may be interdisciplinary or within one specific area of interest. Manuscripts are invited for publication in the following fields of study:

1- Clinical Medicine;
2- Biomedical and Life Sciences;
3- Engineering and Physical Sciences;
4- Mathematics, Statistics and Computer Science;
5- Business, Economics and Law;
6- Human, Behavioral and Social Sciences.

Manuscripts submitted to the series are peer reviewed for scientific rigor followed by the usual Springer standards of editing, production, marketing and distribution. The series will allow authors to showcase their research within the context of a dynamic multidisciplinary platform. The series is open to academics from the University of Palermo but also from other universities around the world. Both scientific and teaching contributions are welcome in this series. The editorial products are addressed to researchers and students and will be published in the English language.

The volumes of the series are single-blind peer-reviewed.

Book proposals can be submitted to the *UNIPA Springer Series Technical Secretariat* at unipaspringer@unipa.it

At the following link, you can find some specific information about the submission procedure to the Editorial Board: https://www.unipa.it/strutture/springer/

More information about this series at https://link.springer.com/bookseries/13175

Francesco Di Paola · Andrea Mercurio

Parametric Experiments in Architecture

A Connection Joint Design for Sustainable Structures in Bamboo

UNIVERSITÀ
DEGLI STUDI
DI PALERMO

Springer

Francesco Di Paola [iD]
DARCH
University of Palermo
Palermo, Italy

Andrea Mercurio
Rozzano, Italy

ISSN 2366-7516 ISSN 2366-7524 (electronic)
UNIPA Springer Series
ISBN 978-3-030-96278-4 ISBN 978-3-030-96276-0 (eBook)
https://doi.org/10.1007/978-3-030-96276-0

This Springer imprint is published by the registered company Springer Nature Switzerland AG
The registered company address is: Gewerbestrasse 11, 6330 Cham, Switzerland

Springer Nature More Media App

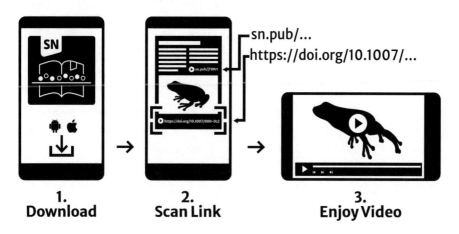

sn.pub/...
https://doi.org/10.1007/...

1.
Download

2.
Scan Link

3.
Enjoy Video

Support: customerservice@springernature.com

Foreword

Digital and digitized, on these two terms still occurs today one of the most frequent misunderstandings in the use of computer aids in every production activity. Naturally, this also happens in the design activity, from the large scale of urban intervention to the minute scale of industrial product. Although it is difficult to trace distinctive boundaries to categorize belonging to one category or the other, between those defined by the two terms indicated, in a general and synthetic way, we could say that a work is digital if the integration and processing of the information are vital for its existence; conversely, a work is digitized if the same ingredients were used simply at reproducing, enhancing and speeding up tools that already belonged to the pre-digital era.

In the case of the digitized, production takes place according to operating methods established in analogical procedure: there is no real innovation, if not just marginal, and the process uses only greater strength and production speed, offered by "brutally" computational aspects. In the case of digital, the methodological process, which leads to the production of the work, is completely innovative, and the computer aid is used to its full potential.

Specifically in the drawing, understood in its meaning as a fundamental aid to the ideation, this duality also travels together with the two modes of operating directly and indirectly on the models that contribute to defining the project. A distinction is anything but trivial, since it affects one of the existential aspects of the designer's work and which concerns the nature and location of the design model.

In direct drawing, mostly "digitized", the model is recursively refined, taking shape first in the mind of the designer and subsequently in the digital space of representation. In indirect drawing, mainly "digital", the designer can only vaguely imagine the shape of the project, since he does not have enough elements to represent it; however, he knows the rules of its evolutionary genesis which, once implemented and digitally processed, will give a definite shape to the designed model.

Within this scenario, Francesco di Paola and Andrea Mercurio show that they are perfectly at ease in codifying and illustrating an operational methodology useful for using bamboo in engineering and architecture projects, a material that is once again in the limelight for its performance and sustainability characteristics.

In fact, this book collects the theoretical foundations and experimental methodologies of bamboo, which originated in an analogue world, then transformed into a
digitized one, but which the authors innovate and project into a more current digital
world.

Precisely, the principles of this digital world are described in the chapter
"Algorithmic-Generative Architecture", in which particular emphasis is given to
the reasons for parametric modelling and to the illustration of the fundamentals of
algorithmic and computational modelling.

In reading what the authors have written, the key element that philologically
connects each of the process methodologies illustrated is evident and which must be
identified in the explanation of the rule underlying the design thought.

A kind of digital, algorithmic *Ekphrasis*, with the aim of optimizing results.
An optimized form design process must rely on interdisciplinary contributions. A
methodological approach that confirms and consolidates the existence of a new operational scenario, matured in the transition between analogue and digital, in which
representation is not a "synthesis" of a quality of the project, but is an expression
of "complexity", understood as an integrated set of qualities that satisfy the design
objectives.

It is therefore no wonder that digital representation is expressed today up to the
construction of the work itself: "representation" in the form of concrete material,
realized automatically, thanks to the multiple manufacturing technologies. A point
of view, known to the authors, who look clearly at the current transformation of the
construction site in the version called "4.0", alluding to the analogous process of
industrial innovation.

After framing and contextualizing the operational area from a methodological and
experimental point of view, the authors introduce the potential and peculiarities of
the building material under study in "Bamboo as building material". Here the factors
of interest and performance of this natural material are retraced and substantiated
by reference data: the history and tradition of its use which certifies its reliability;
sustainability and production economy; the possibility of doing so grows in very
different geographical situations; the reduced need for preparatory treatments for
use and the remarkable mechanical characteristics.

Once the state of the art has been completed, concerning both the scientific operational criteria and the object of study, the authors get to the heart of the experimental
hypothesis and the design of innovative solutions. The path proceeds from the particular to the general, from the structural node to the polyhedral pattern necessary to
approximate a *free-form* surface.

The design of structural node is addressed in "Algorithmic modelling and prototyping of a connection joint for reticular space structures". Through the use of generative and parametric modelling, the authors identify algorithmic procedures, capable
of solving and validating any spatial configuration, of the rods that make up the reticular structure. The shape and size of the elements making up the node are optimized
on the basis of the returned data produced by the behavioural analysis with respect
to the stresses. This first experimental design phase is carried out up to the physical

prototyping of the node, described in great detail and made in PLA: therefore, it is oriented to the first functional geometric checks.

Having solved the particular problem of "how to connect the rods", the authors design, experiment and generalize a useful solution to discretize generic *free-form* surfaces in a set of rods and nodes, configuring a solid polyhedral reticular structure. The solution addressed in "Structural patterns for *free-form* surfaces" has as its origin a mathematical surface, defined in the form of *NURBS*, on which tessellation and subdivision algorithms are applied that are calibrated in order to produce—dimensionally and structurally optimized—a set of bamboo elements that can be made and solidly assembled.

The authors illustrate in great detail the transformation from the mathematical continuity of the free surface to the discrete one of the corresponding polyhedral constructed surfaces: a passage that is rich in complexity and criticality and which, in everyday construction practice, is frequently overly simplified, without fully understanding how much a too reductive approach ends up by significantly affecting the formal quality and optimization of the final construction of the work.

In the theoretical, methodological and experimental experience conducted by the authors, a propositive and further planning action could not be absent which, if on the one hand it concludes the book, on the other it raises the challenge. In "Different Possibilities of Experimentation Design", Francesco Di Paola and Andrea Mercurio, after having clarified the generality of their solutions, applicable also to other materials, illustrate three possible configurations, designed for the construction of a multifunctional pavilion, hypothesized to support the exhibition and relational activities of the teachers and students of the Department of Architecture of the University of Palermo. A solution that we hope the authors can soon implement and that actively involves students and teachers in the educational value of its direct production as well as design.

Rome, Italy   Graziano Mario Valenti

Introduction

The study presents the results of an experimental research that explores in depth the design process of a joining system for bamboo lattice structures, proposing new and advanced solutions that guarantee maximum compositional freedom. The 3D printing production of different types of connection joint-type and the architectural–compositional design of a temporary structure, intended as an exhibition pavilion, have made it possible to validate the defined technological solution.

Currently, bamboo, despite being one of the main natural building materials in the field of sustainable architecture, is not widely used due to the lack of adequate connection systems, especially in Europe, where more restrictive regulations are in force.

The stalks of bamboo are natural hollow profiles characterized by high mechanical performance and are a highly renewable low-cost material. They make it possible to create light and resistant structures, but the dimensional heterogeneity and the tendency to longitudinal splitting of the ends of the stalks make it difficult to make connections capable of fully exploiting the resistance of the material.

Bamboo is a rapidly growing, naturally available renewable resource that is quite strong and lends itself to structural applications. The versatility of use of the material can offer new solutions to cultural heritage conservation and development, especially in poorer geographic areas.

In compliance with the aims of the new *European Green Deal*, that takes up the inspiring principles of the historical Bauhaus movement, founded by the architect Walter Gropius in 1919 in Weimar, the wise use of natural raw materials such as bamboo activates and promotes new design approaches aimed at sustainability and reduced climate impact.

In 2018, UNESCO organized a session at *BARC 2018* "Linking bamboo and World Heritage through creativity", which brings together stakeholders from ministries, universities, world heritage sites and other sectors to discuss how creativity can help promote the conservation of the world's heritage and the sustainable development of local livelihoods.

The *International Bamboo and Rattan Organization* (INBAR) promotes sustainable development strategies from an environmental standpoint using bamboo at the international level with the aim of determining solutions that make it possible to activate political actions, improve regulatory frameworks and define standards and certification schemes governing the use of bamboo as an important resource. As part of the Development Programme, INBAR has developed a digital platform, "INBAR Bamboo Survey", to collect georeferenced data with the aim of improving local management and ensuring easier accessibility to bamboo forests in China and East Africa. This updated map of the largest plantations would allow industries to identify suitable suppliers for reasoned bamboo use and assess the potential for broader development of this growing sector.

Furthermore, today the economic system of the circular economy promotes a different conception of production and consumption, focusing more on extending the life of products, on the production of long-lasting goods and on the reduction of waste. The product is intended as an assembly of biological and technical components, perfectly inserted into a cycle of materials. A virtuous product is designed for disassembly and repurposing, without producing waste. This new and environmentally sustainable approach makes bamboo a very precious resource.

The employment of the material for environmentally conscious and sustainable construction would meet 6 of the 17 *Sustainable Development Goals* (SDGs), "global goals", of the UN 2030 Agenda. The policy document considers in a balanced approach the three dimensions of sustainable development, economic, social and ecological. This has led to a re-evaluation of bamboo, promoted by internationally renowned architects, who have highlighted its potential, in particular the physical and mechanical properties of the stalks, the growth resistance in any type of environment without need for pesticides, the high degree of renewability of the raw material and the versatility in uses dictated by different design, geographical and cultural needs.

The rediscovery of this extraordinary construction material must go hand in hand with the development of IT tools and methods to make it possible to enhance the engineering design and production processes.

Virtual representation, free-form surface modelling techniques and numerical control manufacturing, with their intrinsic dynamic and interactive capabilities, have profoundly expanded and enriched the repertoire of geometric shapes, generating innovative design skills and creative languages.

The creations of contemporary architecture by Peter Eisenman, Zaha Hadid, Norman Foster, Massimiliano Fuksas, Renzo Piano, Toyo Ito or Frank Owen Gehry express unprecedented contributions and experiments in the use of the computer at the service of design and technical tasks and the simultaneous management of form, construction technology and production of all building components.

There is no doubt about the opportunities for exploration, contamination, relationships and overlapping of ideas, measurements and information, which the continuous evolution of expeditious, parametric and automatic procedures brings to the use of the many products of the information age. In this fervent experimentation, the progress of tools supporting the representation culture pushes professionals to a specialized level of knowledge of design techniques, revealing increasingly stimulating fields of application.

However, the intrusiveness of IT development could, at times, create a kind of catharsis with respect to traditional techniques. Operations of graphic synthesis, reduction of scale, understanding and reading of reality, for a subsequent elaboration according to the codes of design, imply the acquisition and mastery of all the methods of the science of representation, especially the traditional ones. It is essential to understand the need to plan the design and to acquire a working practice supported by a method that clearly identifies objectives and purposes.

The added value of digital culture is rooted in the complementarity and synergy of all graphical and expressive methods of architectural language, hinging on the foundations of scientific representation. The latter play a basic role for infographics, constitute an essential cultural baggage and enrich the researcher with the awareness of possessing the tools of knowledge and governance of the geometric properties that regulate space, in order to be able to both read and communicate the design.

Based on these considerations, the text describes an innovative proposal for a connection joint prototype different from those currently known and defines a design strategy for a digital model with a geometric-spatial shape that is parametrically adaptable to any spatial configuration of bamboo rods with heterogeneous dimensions. The described algorithmic definitions manage the design flow as a seamless and integrated process, which goes from the concept to the implementation phases with numerical control machines.

The text examines some possible architectural applications with a particular focus on free-form surfaces, for which the flexibility of the connection system for non-standardisable solutions in sustainable and ecological design is of primary importance.

The current techniques of parametric modelling, the "form-finding" methodology and the optimisation processes with genetic algorithms have constituted the geometric-formal and structural control tools of free-form surfaces, of which the most relevant aspect was that of the discretisation in polygonal meshes that make it possible to define the construction elements of the surface itself.

The design of complex and organic geometric shapes, through the visual programming of digital algorithms (generative, algorithmic and computational modelling), in addition to bringing about a methodological and applicative renewal, has initiated trans-disciplinary investigations.

In industrial design, as in multi-scale architectural design, the explication of algorithmic thinking promotes research directions based on the centrality of the concept of code procedure for building geometric-informative models. The parametric and semantic digital three-dimensional model simulates, collects and manages not only geometric data, but also structural, energy related and construction aspects of the work, putting them in relation with each other and thus improving the interaction and dialogue between the design figures involved in the process.

Furthermore, in the field of design, generative and pre-figurative systems are now often associated with new production processes that are no longer of a "mechanical" type (cutting, turning, milling) but "plastic", linked to the additive modes of *digital fabrication*. In the near future, most industrial processes will have a digital matrix as a generator of governance and production control. The generative approach is useful for the designer to translate even the most complex visions into tangible signs, conceiving objects that can significantly adhere to the specific needs of people, contributing to the construction of unprecedented and fruitful design paths.

Moving in this direction, the text is divided into five chapters; the first two, preparatory to the experimental study, introduce the topic by describing the main theoretical, cognitive and technological aspects, exploring in-depth research experiences in the specific context of the project; the following three chapters describe in stages the methodological process of investigation and its results.

In Chap. 1, "Algorithmic-Generative Architecture", the meaning and evolution of the term "parametric" are briefly described with the different meanings associated with the design process and the purely geometric-formal approach. It describes form-finding simulation and latest generation digital manufacturing techniques, preparatory to the study and design of reticular spatial structures in bamboo, described in the following chapters. It concludes by focusing the attention on some recent experiments of exhibition pavilions, temporary architectural structures that lend themselves well to possible applications of bamboo with new design and construction approaches.

Chapter 2, "Bamboo as Building Material" describes the chemical, physical and mechanical characteristics of bamboo; the main protective treatments that are used to ensure its durability over time against atmospheric agents; the regulatory framework governing its use and the international and national market. The authors are aware of the fact that only continuous research and experimentation on the material will enable the acquisition of the necessary knowledge for its and, consequently, the establishment of consolidated construction practices. In this regard, particular attention is paid to the treatment of the different types of connection systems of bamboo stalks, from the most ancient to the latest generation and innovative solutions, available and described in the literature.

The last paragraph is dedicated to the presentation of two specific species of bamboo: *Bambusa Vulgaris* and *Phyllostachys Viridis*, which grow luxuriantly in the Botanical Garden of the University of Palermo. The supply of a few linear metres of stalks of these two species constituted the raw material of the experimental prototypes.

Chapter 3, "Algorithmic Modelling and Prototyping of a Connection Joint for Reticular Space Structures", describes the detailed investigation process that defines the new and advanced product design solutions of the single element joint, parametrically determining the geometric shape so that it is adaptable to any spatial configuration of the lattice of rods that constitute the structure as a whole. The system enables the junction of a variable number of stalks of heterogeneous dimensions and oriented according to generic directions, guaranteeing maximum compositional freedom to the designer. The algorithmic definitions, processed within the plug-in *Grasshopper* of the well-known modelling software *NURBS Rhinoceros*, made it possible to manage the design flow as an integrated process that goes from the concept to the implementation phases with numerical control machines. Structuring a design process in algorithmic terms has had the advantage of obtaining flexible solutions to the different design conditions according to "form-finding" methods, which can be optimized by means of genetic algorithms.

At the end of the digital process, a description is provided of the full-scale production phases of a spatial reticular structure module with different types of joints and rods made of bamboo stalks, using professional machines for additive rapid prototyping.

Chapter 4, "Structural Patterns for Free-Form Surfaces", analyses a series of topics related to the design and optimisation of a free-form structure. Having determined the shape of a NURBS mathematical surface, we investigate the main subdivision techniques into discrete elements with dimensions suitable for fabrication and assembly. The discretisation method of a planar surface based on the diagrams of Voronoi is combined with "form-finding" and optimization processes in order to meet structural requirements (the form-finding technique is the *Particle-Spring System*, via the plug-in *Kangaroo* for *Grasshopper*, developed by Daniel Piker).

In conclusion, the diagram of the form found structure is optimized by iterating an algorithmic process for determining the best geometric-spatial configuration that meets certain threshold values, through the use of the evolutionary genetic solver of *Grasshopper*, *Galapagos*.

Chapter 5, "Different Possibilities of Experimentation Design", with the aim of showing some of the possible uses of the technological solution, presents three different architectural applications: a roof with a double-layer spatial grid, a geodesic dome and a graphic simulation of a multifunctional free-form pavilion of a temporary nature. The design idea of the multifunctional pavilion aims at making it a meeting place for students, professors and external interlocutors who can use the structure of the pavilion as a place for conferences and workshops, the exhibition of works produced by students and a shared study environment.

Contents

Chapter 1
Algorithmic-Generative Architecture

Abstract The chapter briefly describes the meaning and evolution of the term "parametric" with the different meanings associated with the design process and the purely geometric-formal perspective. In Industrial Design, as in multi-scale architectural design, the explication of algorithmic thinking promotes research directions based on the centrality of the concept of code-procedure for building geometric-informative models. The representation of complex geometric shapes, through the visual programming of digital algorithms (generative, algorithmic and computational modelling), in addition to bringing about a methodological and applicative renewal, has made it possible to initiate trans-disciplinary investigations. The semantic and digital three-dimensional model simulates, collects and manages not only geometric data, but also structural, energy related and construction aspects of an artefact, putting them in relation with each other and thus improving the interaction and dialogue between the design figures involved in the process. *Form-finding* simulation and latest generation digital manufacturing techniques are described, preparatory to the study and design of bamboo structures, described in the following chapters. The chapter concludes by focusing the attention on some recent experiments of exhibition pavilions, temporary architectural structures that lend themselves well to possible applications of bamboo with new design and construction approaches.

1.1 Algorithmic Modelling and Digital Simulation Techniques

The definition of "parametric architecture" is to be attributed to Luigi Moretti, who brings the theme of the architectural project back to the identification and analysis of parameters, which can be expressed mathematically, and of the relationships that exist between them [1].

Moretti identifies as parameters all those design variables that the architect must consider, and to which he or she must respond, in order to meet the functional needs and requirements of the project [2]. The variability of the parameters constitutes the domain in which the freedom of formal decision is exercised. His studies, starting from the end of the 1930s, have the aim of giving a foundation, as objective as

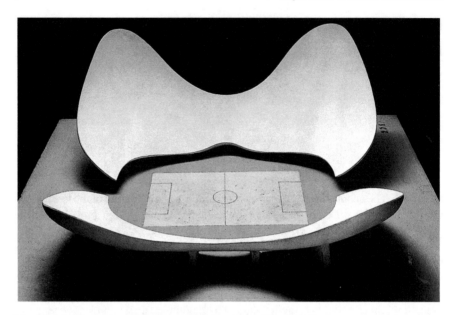

Fig. 1.1 Model of a football stadium. Reproduced from Rostagni [1]

possible, to the design process of even very complex shapes. Reference is made, for example, to the model of a stadium, presented on the occasion of the twelfth triennial of Milan in 1960, whose design takes into account 19 parameters, ranging from the viewing angle of the spectators to the costs of concrete (Fig. 1.1). Moretti sensed well in advance the potential and the role that computers could have in the architectural field, especially in solving parametric problems.

Ivan Sutherland, in 1963, conceived *Sketchpad*, the first CAD program, which allowed drawing, with a pen equipped with a laser pointer, elementary geometric entities, but above all, it made it possible to create associative links and constraints between them, using parametric equations [3]. The program was able to recalculate and redesign simple geometries as the assigned parameters changed [4].

Since the early 1990s, the development of parametric modelling has profoundly influenced the thought process, opening new perspectives for the development of architecture. According to some contemporary theorists, such as Patrik Schumacher, parametric modelling techniques are not only new tools available to designers, but they have led to the birth of a truly global style, a new aesthetic-functional approach called "Parametricism" [5].

In just a few years, the potential of the "parametric" approach has been at the centre of the architectural debate, detailed in particular in a special issue of the magazine *Architectural Design* (2016) entitled *Parametricism 2.0: Rethinking Architecture's Agenda for the 21st Century*, edited by Schumacher himself.

The novelty of this approach, however, does not so much concern the formal and aesthetic outcomes, but rather the logical process that constitutes the sub-structure of the project. From an additive method, in which we proceed by superimposing independent signs on the drawing sheet, we move on to an associative process, in which each element is related to the others.

The transition from drawing on paper to CAD has simply translated the "additive logic" into a digital tool, without conceptually modifying the processing schemes and methods [6]. The parametric software programs define a different approach to the generative design process and the representation of complex geometric shapes, optimising them for use in multiple application fields (structures, energy, economics, etc.). They make it possible to organise projects into associative systems, in which the interrelation between the parts, established through the definition of an algorithm, allows the overall configuration to be altered by propagating the changes at different scales. The difficulty in using such a set of tools is not so much learning a programming language, but rather the ability to correctly formulate a problem in parametric terms.

The field of investigation of this study is precisely linked to the exploration of the potential of current parametric techniques in an integrated and seamless experimental process, ranging from design to manufacture.

An algorithm can be defined as a procedure useful for finding a solution to a given problem, through a sequence of well-defined and unambiguous elementary instructions.

The breaking down of a complex problem, in a logical sequence of elementary operations, is a mechanism close to our way of thinking; moreover, during the elaboration of a project, especially in the conceptual phase, we do not think in terms of form, but of connections and relationships between the parts. From these considerations it is possible to affirm that algorithmic modelling proves to be a natural transposition of a reasoning process into a digital scheme.

Rather than acting directly on the design of the single object, this approach gives the user the possibility to define a logical process, valid for the same family of problems.

Each algorithm requires a finite number of input data or parameters, of a geometric, mathematical or textual type, and gives as a result a unique output, which in the context of the study, will be a geometry (Fig. 1.2) [7].

In relation to the solution they return, there are different types of algorithms: decision-making, if an affirmative or negative answer is returned as output; computational, if numerical quantities are returned; research related, for the determination of solutions; optimisation related, if the need is to determine the best possible solution based on selection criteria.

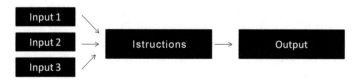

Fig. 1.2 Schematic flowchart of an algorithm

Indeed, through the development of an algorithm, the designer-programmer builds the genetic code, the sub-structure, which governs the entire process. The algorithm, understood as a system of relations and connections, guarantees consistency between the parts and compliance with an overall logic: the modification of the input, which we will more aptly call *parameters*, spreads throughout the system taking into account the relationships that exist within the same.

The tool we have chosen for the generation of algorithms is *Grasshopper*, a plug-in developed since 2007 by the Finnish programmer David Rutten for the modelling software *Rhinoceros* by *Robert McNeel & Associates*. The input and the instructions are provided to the computer through a specific programming language or, in the case of Grasshopper, through a visual scripting editor, while the output of the algorithm is displayed inside the window of the modelling software. In this way, by virtually connecting the various components (which can be geometric or primitive entities, mathematical operations and geometric-spatial transformations) with a sequence of commands (associative logic), the nodal diagram is generated that encodes the parametric geometry associated with it. The resulting nodal system tree within the editor (*Grasshopper*) is characterised by data flowing in one direction: each operation is influenced by those that precede it and it is not possible to start loops, unless using specifically created additional components.

The geometry is visible in the three-dimensional modelling environment (*Rhinoceros*) within the workspace, which is managed and controlled by the editor.

In the phases of structuring the algorithm, a fundamental role is played by the identification of a strategic path that leads from the input to the desired output. The steps and the sub-phases of the process primarily depend on our way of thinking and conceiving a complex geometry: the same result can be obtained following different procedures and consequently choosing different types of parameters; the degree of user control over the generated geometry will depend on this choice.

As a direct consequence of the associative logic, it is possible to create conceptual and effective links between the different levels of design study. The modification of a parameter on a larger scale is able to generate a propagation of modifications such as to get to the congruent redefinition of small-scale details. It is possible to hypothesise a direct link between the parameters relating to the general shape of a complex surface and the geometric characteristics of a structural node, all guided by the relational logic defined by the designer.

The rationalisation of the form, the decomposition and development of complex surfaces into flat elements, cease to be "a posteriori" operations but are integrated into the same formal definition process.

By way of example, two possible strategic paths are shown for the definition of a line inclined at 45° with respect to the horizontal axis, having one end coincident with the origin {0; 0} and the other at the coordinate point {1; 1}. The two elementary algorithms show how, although the result obtained is identical, the required parameters and, therefore, the possibilities of geometric control, are different (Fig. 1.3).

In the first case, the line is uniquely determined by the Cartesian coordinates of the two ends, which constitute the only input; in the second case, once the extremity has been fixed at point A = {0; 0} and the length of the line has been established

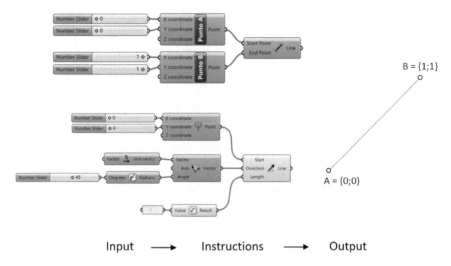

Input ⟶ Instructions ⟶ Output

Fig. 1.3 Algorithmic construction of a line inclined at 45°

as being equal to $\sqrt{2}$, the direction is determined by a vector rotated by 45° with respect to the horizontal plane. Although the first definition appears logically the most immediate for solving this problem, the second algorithm offers the user the possibility to control parametrically and with greater flexibility the length and angle of inclination of the line itself, offering a resolution scheme for a family of problems of the same type.

The number of necessary components is generally inversely proportional to the subsequent degree of control, but the definition of the strategic path and the parameters to be used is a subjective choice that depends on the purposes for which the algorithm was designed. This is a decisive choice, which sometimes require to plan the fundamental steps before moving on to the actual elaboration of the algorithm. It is necessary to envisage and anticipate the possible variants that the project will encounter and consequently structure a sufficiently flexible logical sequence.

One of the greatest potential of this type of modelling is that the designer has the possibility to modify the input parameters at any time, obtaining as a result the possibility to explore infinite variations even in very advanced stages of the design project. Furthermore, of particular interest for the presented study is the possibility of being able to predict, with the use of the plug-in, mechanical behaviours through the simulation of physical forces and, consequently, to be able to identify structural solutions directly in the design phase. As will be described below, one of the strengths of computational design is the topological optimisation based on the performance of predetermined materials.

The following graph shows how the possibilities of making changes to a project generally follow a decreasing trend over time, parallel to a progressive increase in the costs that such variations would entail. This factor emphasises the importance of

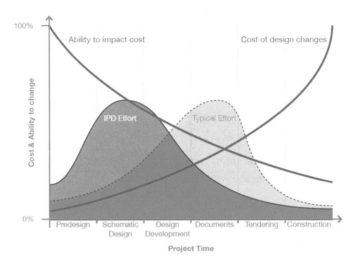

Fig. 1.4 MacLeamy curve. Reproduced from MacLeamy [8]

concentrating investments in the preliminary stages of the design when the costs of the variants are limited (Fig. 1.4) [8].

Samuel Geisberg, founder in 1985 of *Parametric Technology Corporation* (PTC), states that the main goal for which parametric modelling was introduced and developed was to create a flexible system that would allow engineers to consider multiple design alternatives while minimising the costs of variants [9]. The latter are sometimes predictable with careful planning, but are often due to external factors such as legislative changes, customer choices or changes in material costs.

Therefore, in addition to shifting the economic burden and time investments towards the preliminary stages of design, a further strategy is to bring the costs due to variants as close to zero as possible through parametric modelling.

Digital technologies, therefore, effectively contribute to the management of information flows and geometric-spatial relationships generated by an architectural project. In the analysis of complex engineering and architectural problems in which different performance variables responding to heterogeneous requirements interact, the use of genetic evolutionary algorithms triggers optimisation processes. Indeed, genetic algorithms can be a flexible tool for exploring different design strategies and find excellent application in complex contexts, which require the evaluation of multiple performance requirements. The use of the algorithm can be extended to the different design phases, from the conception to the exploration of different alternatives and to the final phase of defining specific solutions.

The genetic algorithm was formalised in 1975 by John Henry Holland following research on the adaptation of natural and artificial systems. Genetic algorithms constitute a probabilistic method for optimisation. Their underlying principle, natural selection, derives directly from Darwin's theories on the evolution of the species. Nature dictates that in the reproduction of a species the genetic code undergoes random mutations, responsible for the diversity of individuals. Natural selection consists in the survival of the fittest, or more appropriately of the individuals who best adapt to environmental conditions [10].

Genetic algorithmic logic is widely used in logistics and industrial planning (for example in solving planning issues for a transport vehicle fleet, in planning production in industrial plants or in scheduling tasks for instance in the case of an airline).

At the same time, the same logic guides our criteria every time we define a design choice; the steps we follow are the following: the generation of a certain number of starting solutions (*genome*), the selection of some of them, the mutation of the latter and the recombination that makes it possible to obtain a new set of solutions, from which to repeat the procedure until obtaining the solution that we consider to be the best. Therefore, it is an optimisation process whose result depends on our ability to choose the best solution [11].

Starting from the 1950s and 1960s, many programmers understood that genetic algorithms could be used as an optimisation tool for architectural engineering problems, when the search for the best solution among a large population of possible candidates is required [12, 13].

The selection criterion is based on the assignment of a score, defined as *fitness*, to each member of the genome; the fitness depends on the effectiveness of that particular solution in solving the problem.

A popular method of setting the algorithm is to minimise, or maximize, the fitness function $f(x)$ by acting on a set of parameters $(x_1, x_2, ..., x_n)$ that vary within its own domain. Every set of parameters constitutes a member of the starting population. The output provided by the genetic solver is, as can be understood from its probabilistic functioning, an approximate solution, which may not be the absolute best [14]. From time to time, different results can be obtained from the selection of a random starting population [15].

As part of the study, the genetic algorithms will be used in combination with the data obtained from FEM analyses, automatically iterating the process of redesigning and analysing the various geometric-spatial solutions of the joints of a reticular structure in bamboo culms, up to the selection of the most performing solution with respect to the chosen fitness.

1.2   Form-Finding

The term *form-finding* refers to a process that makes it possible to define the geometry of a structure in equilibrium with respect to a given load condition. The geometry is the unknown factor of the form-finding process, but an arbitrary starting geometry is in any case necessary.

Once the final shape has been determined, often using as parameters only the weight of the structure and the mechanical characteristics of the fictitious materials, we proceed by updating the model with the real properties and dimensions of the elements and carrying out, only at a later stage, the actual structural analysis.

The form-finding techniques have been developed in close relationship with structures such as those commonly referred to as *shells* in literature, being resistant by virtue of their shape.

In the design of shell structures, in reinforced concrete, brick or other material, the main objective to be achieved is to transfer the loads to the ground by pure compression, minimising the moments. To this end, physical models have been developed, in which fabrics or ropes are anchored and subjected to the force of gravity, as well as virtual models based on numerical simulations.

The oldest and most widespread method was conceived by Robert Hooke in 1676; the idea is to invert the shape obtained from a rope anchored at the ends, subjected exclusively to traction, in order to obtain an arch in pure compression.

If the weight per unit of length is constant, the shape obtained will take the name of catenary. Galileo Galilei himself confused this shape with that of a parabola, which instead would be obtained if the load were uniformly distributed horizontally.

In most cases, the ratio between the distance L between the support points and the height of the arc d is in the range $2 < L/d < 10$.

The catenary concept can be extended in space to the case of vaults and shells, in which membrane actions are predominant.

Real scale models have long represented a useful tool for evaluating the physical and mechanical behaviour of a shape; this has made it possible to combine formal design and structural calculation in a single process.

The *Sagrada Familia* by Gaudì (1852–1926) in Barcelona is probably the best-known example of architecture whose shape is the result of a form-finding process [16]. Another example is the model for the crypt of the *Colonia Güell* made with hanging ropes and sand packets proportional to the loads (Fig. 1.5).

In 1969, Sergio Musumeci built a bridge over the Basento river in Potenza, the shape of which was obtained through different form-finding techniques. The first models were made with soap films, while the final shape was defined with a fabric model (Fig. 1.6).

Frei Otto has created numerous *gridshells* often using scale models to validate the mechanical behaviour of his structures. This is the case of the *Multihalle* in Mannheim (1973–74), for which models were made at various scales with the collaboration of *Ove Arup & Partners* (Fig. 1.7).

The study of the shape through the use of scale models clearly shows the search for a parametric control on the calculated geometries.

A model offers the possibility to modify some parameters, such as for example the length of the ropes, the position and entity of the loads, the position of the constraints etc., automatically generating new shapes and spatial configurations. The processing of the algorithm, which leads from the incoming data to the output and, therefore, the definition of the form, is entrusted solely to the laws that govern physical reality: Hooke's law and gravity.

The exploration of the different solutions offered by the model constitutes a powerful parametric tool extensively used by Antoni Gaudì and Frei Otto [17].

In the past, scale models have replaced complex, inaccurate and time-consuming numerical computational methods. However, the structural behaviour of a model

Fig. 1.5 Reproduction of the crypt model of the Colonia Güell by Gaudì, Barcelona, created in 1982–83 by the Institute for Lightweight Structures. Reproduced from Adriaenssens et al. [18]

Fig. 1.6 View of the bridge over the Basento river, Potenza. Reproduced from Tedeschi [7]

Fig. 1.7 View of Frei Otto's Multihalle in Mannheim. Reproduced from Adriaenssens et al. [18]

does not always match what is expected at full scale. Galileo Galilei himself noted that, by scaling a cube, some characteristics such as the area vary with the square of the scale factor, others vary linearly, while the mass, for example, which depends on the volume, will be proportional to the cube of the scale factor adopted. Many other factors are strongly influenced by the scale of the model, including the stiffness of the material, the Poisson's ratio and the effects of temperature and humidity [18].

Despite the approximations, it must be emphasised that to this day physical models remain a valid tool for exploring the form and verifying the results obtained from computer media.

In recent decades, with the strong development of digital modelling, the *form-finding* techniques have been digitalised using the calculation capabilities of computers.

Developed methods include *Natural Shape Finding, Force Density Method, Thrust Network Analysis, Dynamic Relaxation and Particle-Spring System*. Some methods are aimed at a particular category of structures as in the case of TNA, developed by Philippe Block's research group at the MIT, for the design and analysis of brick vaults. This is a method based on the procedures of graphic statics applied to three-dimensional systems. Other techniques are distinguished, for example, by the type of elements being analysed, rods or surfaces, or by the type of information to be defined as input.

In this study we will use the *Particle-Spring System* method, by means of the *Kangaroo*[1] plug-in for *Grasshopper*, developed by Daniel Piker.

The main purpose of the *Particle-Spring* method is to achieve, through a series of iterations, a static balance. The physical behaviour of a deformable body is thus simulated, in a way that is completely comparable to the physical models mentioned previously.

A certain surface, or the network of lines that make up the structure, is broken down into particles (points), in which the masses are concentrated, and perfectly elastic springs (lines), which deform according to Hooke's law; each of them is characterised by a rest length and a stiffness value [19].

As for a scale model, also in this case the external constraints will be established, which will not undergo any displacement, and the forces, primarily gravity.

Each particle, subject to the action of the elastic forces of the springs and the force of gravity, will enter into motion according to Newton's second law,

$$F = m \cdot a$$

until the balance of forces in each node is reached, through an iterative and dynamic calculation. Since the particles behave like spherical hinges, the balance solution will be characterised exclusively by axial forces on the rods.

The ability to interface the *form-finding* process with FEM (*Finit Element Method*) analyses, albeit preliminary and approximate, gives the designer the possibility to check the static behaviour of the structure in an interactive and immediate way, ensuring that the choices are made more consciously.

1.3   Digital Fabrication

Digital manufacturing techniques (or *fabbing*), in particular by means of CAD/CAM design software and CNC (Computer Numerical Control) machines, make it possible to create physical objects directly starting from 3D digital geometries and bypassing the production and interpretation of technical drawings, which until today have represented the only communication tool between two phases, design and production, which have always been distinct and were never really integrated into a single process.

Thanks to the growing development of automated technologies that increase the degree of precision while reducing production costs, digital industrial production is used in many sectors, exploring new experimental scenarios that combine complex geometric shapes and new materials.

The development of digital manufacturing techniques, such as 3D printing, CNC milling or laser cutting, has two direct consequences in the construction industry [20]:

---

[1] Physics engine for interactive simulation, optimisation and *form-finding*; available on the website http://www.food4rhino.com/project/kangaroo?etx. Last accessed 25/03/2020.

- on the one hand, the reduction of the necessary tolerances and greater precision of the elements produced, with an ever-smaller difference between the virtual model and the physical object; this enables the management and control of structures, even with very complex geometries, with margins of error close to zero.
- on the other hand, the possibility of creating unique pieces, without increases in production costs.

This last statement appears to be in contradiction with the laws underlying industrial production of the last two centuries. In the business logic, the goal of reducing unit production costs was achieved through the standardisation and serial production of identical elements.

In the case of numerically controlled machines, however, the fact that the elements are the same or different from each other is completely irrelevant, since the machine does not know, for example, the difference between a straight line and a curved line; for cutting, it will simply follow a series of spatial coordinates provided as input.

Let us now consider the case of a cube produced through 3D printing techniques and compare a regular cube with a cube deformed by a rotation around its vertical axis (Fig. 1.8).

The printing times, the quantity of material used and the level of detail, as well as the production costs are the same and are not influenced by the form of the object, except in special cases that require specific techniques or precautions.

The difference between the two objects is rather of an aesthetic nature, but the apparent geometric complexity of the artefact does not actually correspond to a real production complexity.

Among the numerical control manufacturing techniques, it is possible to distinguish two macro-categories. The subtractive CNC techniques work by removing material (by digging or cutting), making it possible to create a prototype as a sculptor would do. Through successive and often complex working phases that use milling machines, punching machines, bending machines, lathes or laser machines, material is removed from a solid and full block.

Fig. 1.8 Apparent production difficulty. Reproduced from Davis [4]

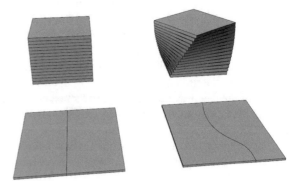

The additive techniques (AM), more widespread today, create three-dimensional objects from the superimposition of successive layers of material (layer manufacturing), obtaining complex geometries with unique processes often impossible to replicate with subtractive manufacturing (Figs. 1.9 and 1.10).

The group ASTM (American Society for Testing and Materials) "Global Additive Manufacturing Programs" has developed a series of international standards that classify additive manufacturing processes into 7 categories [21, 22].

The 7 processes that make possible the manufacture of prototypes, through the use of various types of materials, generally thermoplastic resins, photopolymers or sintered powders, but also metals, ceramics, clays, sands, gypsum or food material, are classified as follows.

- Vat Photopolymerisation:

  - Stereolithography (SLA);
  - Digital Light Processing (DLP);
  - Continuous Liquid Interface Production (CLIP).

- Material Jetting
- Binder Jetting
- Material Extrusion:

Fig. 1.9 Different types of 3D digital manufacturing technologies and processes, subtractive techniques (American Society for Testing and Materials, ASTM)

Fig. 1.10 Different types of 3D digital manufacturing technologies and processes, additive techniques (American Society for Testing and Materials, ASTM)

- – Fused Deposition Modelling (FDM);
- – Fused Filament Fabrication (FFF).

- • Powder Bed Fusion:

  - – Selective laser sintering (SLS)

- • Sheet Lamination.
- • Directed Energy Deposition.

The consolidated 3D printing techniques to produce prototypes, known for decades, are: Stereolithography (SLA); laser sintering (SLS) and Fused Deposition Modelling (FDM).

Stereolithography was the first technique to be used for rapid prototyping. A laser beam, directed by a system of mirrors, hits a photosensitive liquid resin, contained inside a tank, causing it to polymerase. The platform on which the piece rests translate vertically, making it possible to proceed layer by layer. The subsequent treatment inside a UV oven guarantees the resin's perfect solidification.

Laser sintering enables rapid prototyping through the selective melting of powders of thermoplastic materials, metals, ceramics, etc. by means of a laser beam. The powder is contained in a tank, and a laying platform moves downwards at discrete

Δz intervals; at the same time as the lowering of the platform, powder is added and the smoothing of the surface occurs by means of a roller. This technique does not require additional support material, as the powder contained in the tank guarantees the stability of the object.

The well-known rapid prototyping (RP) techniques, more commonly known as "3D printing", work with the AM method and fall into the process category called Fused Deposition Modelling.

Consumer technology makes it possible to create physical objects starting from digital models through an additive manufacturing process, in which the material, a thermoplastic polymer (usually ABS or PLA), is deposited layer by layer until obtaining the complete object [23].

The plastic filament passes through a heated tube and is immediately melted by a resistor before being extruded through a nozzle. Once deposited, the polymer wire immediately solidifies adhering to the rest of the material.

1.4   Parametric Experiments in Temporary Architectures

The possibilities offered by digital tools have made it possible in recent years to extend the use of the computer from a tool aimed at speeding up repetitive operations to a tool for control and formal experimentation. Parametric software, formerly used only by large design firms, have now become accessible to architects, engineers and designers, thanks to the development of interfaces that do not require the knowledge of specific programming languages.

The construction of temporary structures has represented one of the major areas of experimentation in the implementation of parametric techniques. The flexibility of the functional requirements needed for this type of structures make them an opportunity to meet theoretical research and production methods.

Among the recent application examples of this type of approach there are numerous pavilions created in collaboration with the academic world. We mention some of the most representative ones. Between 2005 and 2009, the *Architectural Association Hooke Park*, in Dorset (England), within the program *AA Design & Make*, created a series of pavilions using parametric computational techniques and wood as the main building material (Fig. 1.11) [24].

We should also mention the *ITKE Research Pavilion*, created in 2011 as the result of the collaboration between the *Institute for Computational Design and Construction* (ICD) and the *Institute of Building Structures and Structural Design* (ITKE) of the University of Stuttgart, Germany. It is a structure in plywood sheets with a thickness of 6.5 mm composed of over 850 panels of different geometry made with numerical control machines (CNC) (Fig. 1.12) [25]. In 2019, the University of Stuttgart will establish its new *Cluster of Excellence on Integrative Computational Design and Construction for Architecture*. For the first time, a *Cluster of Excellence* was assigned to the Architecture sector. It will contribute to the creation of an internationally visible research centre with the ambitious goal of harnessing the full potential of

Fig. 1.11 Swoosh Pavilion, London, England 2007/08. Reproduced from Self et al. [24]

Fig. 1.12 ITKE Research Pavilion, Stuttgart, Germany, 2011. Reproduced from Sack [25]

digital technologies to rethink design, manufacturing and construction in the building sector.

Although it cannot be considered a temporary architecture, the *Metropol Parasol*, designed by *J. Mayer H. Architects*, which immediately became an icon of the city

of Seville, is attributable to the world of parametric architecture and digital manufacturing. The structural project, entrusted to the international engineering company
Arup, required the development of a parametric model in which, through automated
iterative processes, it was possible to determine the thickness of each wooden panel
and optimise the connection details in each of the 3000 intersections. The wooden
panels have been CNC-cut from large rectangular panels and arranged in such a way
as to minimise material waste (Fig. 1.13) [26].

In May 2019, researcher Martin Alvarez from the *Pratt Institute*, New York
(USA) dealing with research and robotics, presented the new *BuGa Wood Pavilion*
made in Bundesgartenschau, Heilbronn (Germany). The unique shell construction
designed in the laboratories of the University of Stuttgart represents a new approach
to digital construction in lightweight wood. During the design phase of the structure, a robotic production platform was developed for the automated assembly and
milling of 376 hollow wood segments. This manufacturing process ensured that all
segments fit together with sub-millimeter precision like a large three-dimensional
puzzle (Fig. 1.14).

The common intent of this type of architecture is to create unique structures, overcoming the logic of standardisation in favour of the optimisation and differentiation
of components. Numerical control machines, associated with a parametric design,
ideally enable the direct passage from the virtual model to the real object.

Fig. 1.13 Metropol Parasol, Seville, Spain, 2010. Reproduced from Pohl [26]

Fig. 1.14 BuGa Wood Pavilion, Bundesgartenschau Heilbronn, Germany, 2019. Reproduced from Menges [27]

References

1. Rostagni C (2008) Luigi Moretti 1907–1973. Mondadori Electa, Milano
2. Moretti L (1951) Struttura come forma. Spazio, Roma
3. Sutherland I. (1963) Sketchpad: a man-machine graphical communication system. PhD dissertation, Massachusetts Institute of Technology
4. Davis D (2013) Modelled on software engineering: flexible parametric models in the practice of architecture. Degree dissertation, RMIT University
5. Schumacher P (2009) Parametricism: a new global style for architecture and urban design. In: Digital cities AD: architectural design, vol 79. Wiley
6. Pugnale A (2012) Engineering Architecture: come il virtuale si fa reale. Bloom 14:17–24
7. Tedeschi A (2014) AAD algorithms-aided design. Parametric strategies using Grasshopper. Le Penseur Publisher, Potenza
8. MacLeamy P (2010) Bim-Bam-Boom! How to build greener. High-performance Buildings, HOK Renew
9. Teresko J (1993) Parametric technology corporation: changing the way products are designed. Industry Week Magazine, 20 Dec
10. Darwin C (2010) L'origine della specie. Newton Compton Editori (Ist edn. 1859), Roma
11. Milos D (2011) Structural optimization of grid shells based on genetic algorithms. PhD dissertation, University of Stuttgart

12. Mitchell M (1996) An introduction to genetic algorithms. MIT Press, London
13. Caldas L, Norford L (1999) A genetic algorithm tool for design optimization. In: Proceedings of ACADIA
14. Statopoulos N, Weinstock M (2006) Advanced simulation in design, in techniques and technologies in morphogenetic design. In: AD architectural design. Wiley, London
15. Luebkeman K, Shea K (2005) Computational design+optimization in building practice. Arup J 3
16. Pugnale A (2009) Engineering architecture. Advances of a technological practice. PhD dissertation, Politecnico di Torino
17. Davis D (2013) A history of parametric. http://www.danieldavis.com/a-history-of-parametric/. Accessed 24 Mar 2020
18. Adriaenssens S, Block P, Veenendaal D, Williams C (2014) Shell structures for architecture: form-finding and optimization. Routledge, New York
19. Block P (2009) Thrust network analysis. Exploring three-dimensional equilibrium. PhD dissertation, Massachusetts Institute of Technology
20. Parsons M (2014) Tolerance and customisation: a question of value. Australian Design Review. http://www.australiandesignreview.com/architecture/41321-tolerance-and-customisation-a-question-of-value. Accessed 24 Mar 2020
21. Bird J (2012) Exploring the 3D printing opportunity. The Financial Times. https://www.ft.com/content/6dc11070-d763-11e1-a378-00144feabdc0. Accessed 10 Feb 2021
22. Lu L, Sharf A, Zhao H, Wei Y, Fan Q, Chen X, Savoye Y, Tu C, Cohen-Or D, Chen B (2014) Build-to-last: strength to weight 3D printed objects. ACM Trans Graphics (Proc SIGGRAPH) 33(4):97:1–97:10
23. Excell J (2013) The rise of additive manufacturing. The Engineer
24. Self M, Walker C (2010) AA agendas 9: making pavilions. AA Print Studio, London
25. Sack F (2016) ICD/ITKE Research Pavilion 2011. Open House 2, Jovis Verlag GmbH, Berlin, p 20. ISBN: 978-3-86859-393-8
26. Pohl EB (2011) Waffle urbanism. Domus 947. https://www.domusweb.it/en/architecture/2011/05/10/waffle-urbanism.html. Accessed 10 Feb 2021
27. Menges A (2019) BUGA Wood Pavilion. http://www.achimmenges.net/?p=20987. Accessed 10 Feb 2021

Chapter 2
Bamboo as Building Material

Abstract "Bamboo as building material" describes: the chemical, physical and mechanical characteristics of bamboo; the main protective treatments that are used to ensure its durability over time against atmospheric agents; the regulatory framework governing its use and the international and national market. The authors know only continuous research and experimentation on the material will enable the acquisition of the necessary knowledge for its use and, consequently, the establishment of consolidated construction practices. In this regard, particular attention is paid to the treatment of the different types of connection systems of bamboo stalks, from the most ancient to the latest generation and innovative solutions, available and described in the literature. The last paragraph is dedicated to the presentation of two specific species of bamboo: *Bambusa vulgaris* and *Phyllostachys viridis*, which grow luxuriantly in the Botanical Garden of the University of Palermo. The supply of a few linear meters of stalks of these two species constituted the raw material of the experimental prototypes.

2.1 The Bamboo Potentiality

Bamboo is one of the oldest building materials, having been traditionally used in regions where the plant naturally grows abundantly, such as South America, Africa and, in particular, South-east Asia [1].

Concrete, steel and wood have relegated the use of bamboo to areas that are lacking resources, so much so that it has traditionally been known as "poor men timber" [2]. However, the increased attention towards environmental issues has brought the consumption of natural resources to the forefront; in the construction industry, this has led to a consistent search for sustainable materials, which tend to reduce the energy involved in the production, management and material-disposal phases. Today, the economic system of the circular economy encourages a different idea of production and consumption, focusing more on extending products' shelf lives, producing long-lasting goods and reducing the production of waste. A virtuous product is designed to be taken apart and repurposed, without producing waste.

Table 2.1 Energy needed for production compared with stress at service load [6]

Material	Specific weight (kg/m^3)	Production energy (MJ/kg)	Production energy (MJ/m^3)	Tensile stress at service (N/mm^2)	Production energy (MJ/m^3) to unit tensile stress at service (N/mm^2) ratio
Concrete	2400	0.8	1920	8	240
Steel	7800	30	234,000	160	1500
Wood	600	1	600	7.5	80
Bamboo	600	0.5	300	10	30

This new, sustainable approach makes bamboo an exceptionally valuable resource and has led to a re-evaluation of bamboo, which has been promoted by internationally renowned architects for its potential. This is particularly noteworthy for the physical–mechanical properties of its culm, its ability to grow in any environment without the need for pesticides, its high level of renewability as a raw material and its versatility for different geographical and cultural needs [3].

The recent interest in this latter aspect is evidenced by the need for "*Rapidly Renewable Materials*" parameters in order to obtain LEED (Leadership in Energy and Environmental Design)[1] certification, which requires the use of plants with a life cycle of less than 10 years for construction purposes. Bamboo reaches maturity after only 4–5 years, at which point it can be harvested; it is therefore a highly renewable resource, when we consider that wood requires decades to reach quantities and biomass of similar characteristics (Table 2.1) [4].

Considering the speed of growth and maturation, it has been noted that bamboo is 3–4 times more efficient than wood in terms of annual production [5].

The Dutch architect Janssen [6], who is the founder of the Bamboo Engineering programme within the INBAR (*International Network for Bamboo and Rattan*), carried out an interesting comparison between the environmental impact of the most common building materials and bamboo, taking into account the production energy per unit of each material. The reported values are indicative and are only intended to offer an order of magnitude, but it is clear that, when mechanical performance is equal, the environmental cost of bamboo structures is much lower than structures made from concrete or steel.

The efficiency of the material—which is defined based on the ratio between mechanical performance and specific weight—is comparable to that of steel, hence the use of the name "vegetal steel" to describe bamboo.

In 2005, a study carried out by Pablo Van Der Lugt analysed the environmental costs of building a pedestrian bridge from bamboo in Rotterdam.

---

[1] Energy certification and sustainability programme developed by the Green Building Council in the United States and applied in more than 130 countries; it is the most widely used means for evaluating a building's sustainability in the world. See http://www.usgbc.org/certification [last accessed on 19/03/2019].

To do so, the methodology employed is an LCA (*Life Cycle Analysis*), which is accredited by ISO 14041. It takes all of the environmental effects of a product into consideration, from the production phase through demolition and recycling.

The quantity of material needed based on the strength of the elements, the useful life of the structure's components, the need for yearly maintenance and the related costs for this, the production processes, waste and potential for recycling and transport were assessed for each of the possible alternatives (wood, concrete and steel). Processing all of the inputs made it possible to express the environmental impact of each solution via a single index.

In terms of environmental costs, bamboo was approximately 20 times more sustainable than the other structural alternatives, despite the additional need for transporting the material from Costa Rica by boat (Table 2.2) (Fig. 2.1).

The main reasons for the increased environmental efficiency of bamboo can be largely attributed to two factors: the first is the hollow circular shape of the section, which makes it much more efficient from a structural point of view than a fully rectangular section, as is typical of wooden elements; the second factor, pertains to the production process; bamboo is very productive even without special care and

Table 2.2 Bamboo mechanical properties [6]

Material	Stress at service load (N/mm²)	Elastic modulus (N/mm²)	Specific weight (kg/m³)	Service stress/specific weight ratio	Elastic modulus/specific weight ratio
Concrete	8	5000	2400	0.003	10
Steel	160	210,000	7800	0.020	27
Wood	7.5	11,000	600	0.013	18
Bamboo	10	20,000	600	0.017	33

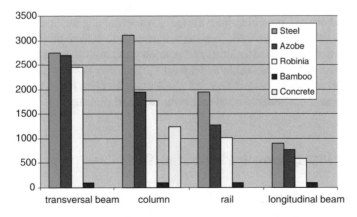

Fig. 2.1 Index of the annual environmental costs (in mPt) for the various elements of the Amsterdamse Bos bridge. The alternative with the smallest value, that is, bamboo, is set as equal to 100 [5]

the transformation into a building materials is limited to maturing the bamboo and applying the appropriate protective treatments.

Furthermore, the purchase costs for the material needed to build the bridge significantly favors the use of bamboo.

In terms of annual costs, steel was found to be most advantageous, thanks to the longer useful life of steel elements. However, bamboo was shown to be a comparable alternative to wood [7].

The conclusions of this study—as well as the advantages—highlight some of the critical points pertaining to the use of bamboo in Europe, compared to more traditional building materials:

- the use of foreign workers increases costs and causes communication problems on the work site, thus leading to delays;
- despite the low cost of importing bamboo, more quality checks and specific treatments are required to avoid the material deteriorating during transport from countries outside Europe;
- due to the absence of a sufficient regulatory framework, additional calculations and tests are required in order to obtain the necessary authorisations from public authorities;
- the shape of the elements (hollow and tapered) and the irregularity of the material;
- lack of appropriate technical expertise on behalf of designers and workers.

It seems clear then that, even today, the major limitations regarding the use of bamboo in Europe are linked to a lack of awareness around the potential uses of this natural material. Since it is a natural material, it has irregularities that make it more suitable for use in construction projects that are more flexible and adaptable in terms of size requirements, such as temporary structures (for example, exhibition pavilions and tents) or small civil projects (for example, bridges and walkways). When it is used correctly, it can be a great substitute for wood, steel and concrete in architectural designs that are aimed at environmental sustainability.

The buildings of Colombian-born Simon Vélez are among the first contemporary structures of this kind in Europe. They make it possible to collect data, static tests and simulations that can be used to define certification standards for the use of bamboo as a building material. His most significant works include the *Zeri* Pavilion, which was built in 2000 for the Hanover Expo and made using bamboo (from the Colombian *Guadua angustifolia* species) combined with steel and cement.

2.2   The Plant, Growth Phases and Their Uses

Bamboo belongs to the *Gramineae* family, which also encompasses cereals like wheat, rice, corn or barley and many plants like the common giant reed (*Arundo Donax*) or wheatgrass. Some of these are annual, as they produce their own seeds every year, while others, including bamboo, are evergreen and have longer flowering cycles, with cadences that can even reach up to every 120 years and more.

The term "Bamboo" comes from the Malay word "Mambu", which was subsequently translated into English as the more commonly known *Bamboo*.

The flowering of a specific species may occur at the same time for all members with a common origin (*gregarious flowering*) and can last for several years, until the plant decays and dies due to the energy expended reproducing. In other species, such as *Guadua angustifolia*, individual plants flower at different times and at irregular intervals (*sporadic flowering*); the plant gets weak but survives and, unlike the earlier example, does not create ecological imbalances due to the death of vast swathes of bamboo.

The temporal extension of the flowering cycles makes botanical classification difficult, as this is often carried out by observing the flowers [8]. There are over 1400 species, divided into two large groups: woody bamboo (*Bambùseae*) with approximately 1290 species and herbaceous bamboo (*Olyreae*) with approximately 115 species.

As has already been mentioned, bamboo has a vast geographic distribution due to its ability to adapt; native to the tropical and sub-tropical regions of the Far East, it grows naturally in Africa, Oceania, Latin America and also adapts to temperate climates. Some species can be found in very harsh climates, up to 4000 m in altitude: *Sasa kurilensis* resists temperatures lower than 20° below zero on the Kuril Islands in Russia, while other species grow in the Himalayas (Fig. 2.2).

Although the recent discovery of bamboo fossil remains in France points to the species' presence in Europe before the great glacial periods, we can say that the plant was introduced to Europe only at the beginning of the nineteenth century (in 1827 in France and 1832 in England). At this time, the advent of steamboats considerably shortened the length of voyages, allowing the transport of green material [9].

The Mediterranean climate of the Costa Tropical is particularly good for planting certain varieties. The most suitable species for temperature climates, in which the minimum temperature does not fall below zero, are those that are typical of tropical and sub-tropical climates. Currently, interesting experiments are being carried out at the University of Granada (Spain), at the Escuela Técnica Superior Ingeniería

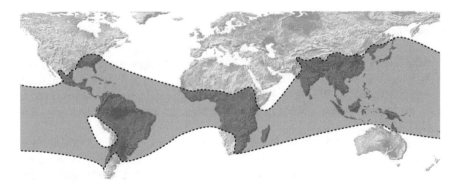

Fig. 2.2 Geographic distribution [1]

Fig. 2.3 Small planting of *Dendrocalamus giganteus* around the central water lily pond of the Botanical Garden of Palermo

de Edificacíon (ADIME), on bamboo belonging to the *Guadua augustifolia Kunth* variety.

The bamboo collection at the Botanical Gardens of Palermo is one of the most extensive in Italy. The English writer Douglas Sladen previously describes the natural splendour of the Palermo Garden in this respect, centring on the description of the bamboo grove around the central fountain. He claims to have seen culms that are six inches in diameter—or approximately 16 cm—which is surely a reference to *Dendrocalamus giganteus* (Fig. 2.3) [10].

Among the species found at the Botanical Gardens of Palermo, it is important to note:

- *Dendrocalamus giganteus*: this large bamboo is of Indian origin and thrives in warm, humid tropical climates. It adapts very well to regions that are temperate, windless and close to the sea, where orange grows. This species is monocarpic, that is, it reaches maturity only once in its life, flowering, producing fruits and seeds and then dying. Its life cycle is approximately 120/160 years. It is considered to be the tallest bamboo, reaching 25–35 m in high and 30 cm in diameter (Fig. 2.4) [11].
- *Bambusa vulgaris*: this bamboo has a dark-green culm and, in the Mediterranean region, reaches 10 m in height and 7–8 cm in diameter. It thrives in humid environments but is adapted to semi-arid regions. It is native to Madagascar and China, where it is used for lightweight structures, houses, huts, boats, scaffolding, furnishing, musical instruments and general crafts (Fig. 2.5).

Fig. 2.4 Culms of the *Dendrocalamus giganteus* species at the Botanical Garden of Palermo

Fig. 2.5 Culms of the *Bambusa vulgaris* species at the Botanical Garden of Palermo

- *Bambusa vulgaris vittata*: this is a varied form of *B. vulgaris* and is also known as "striped" due to its yellow colouring with green stripes. It is especially common in the East, for decorative purposes (Fig. 2.6).
- *Phyllostachys nigra*: this species was introduced in Europe in the first half of the nineteenth century. It is characterised by the colour of its culms, which turns from dark green to pitch black with time. The curved appearance of the culms, which is admired from a decorative point of view, limits its use as a building material (Fig. 2.7).
- *Phyllostachys viridis*: these reach heights greater than 10 m and a diameter of 8–10 cm. They are green in colour, but, over the years and after exposure to sunlight,

Fig. 2.6 Culms of the *Bambusa vulgaris Vittata* species at the Botanical Garden of Palermo

Fig. 2.7 Culms of the *Phyllostachys nigra* species at the Botanical Garden of Palermo

they turn to a golden-yellow hue. There is a white ring beneath each node, which is comprised of a waxy substance known as pruina. The internodes are long and typically regular. The branches, which spring up on the nodes, alternatively on one side or the other of the culm, create a typical groove that disrupts the perfect circularity of the section (Fig. 2.8).

Other bamboo species found here include *Bambusa ventricosa*, *Phyllostachys violascens*, *Phyllostachys bambusoides* and *Otatea acuminata*. Given the ready availability of the bamboo plant, bamboo is used in the construction of bridges,

Fig. 2.8 Culms of the *Phyllostachys viridis* species at the Botanical Garden of Palermo

Table 2.3 Growth phases of bamboo and potential uses [4]

Age	Use
< 30 days	Edible bamboo shoots
6–9 months	Weaving, baskets, decorations
2–3 years	Fibre for laminates and panels
3–6 years	Culms for construction
> 6 years	Physical properties gradually diminish

houses, furniture and laminates in South-east Asia and Latin America, since it can be considered as a fantastic alternative to wood in a wide variety of contexts.

The age at which the culm must be cut depends on the use for which it is intended, since this is a determining factor for the mechanical properties of the building material. At approximately 6–9 months, the plant can be used to make chests and baskets, taking advantage of the flexibility of the young fibre; from 2 to 3 years, it can be used to manufacture panels or laminates; once they reach 3–6 years, the culms are suitable for structural use; after 6 years, the bamboo progressively loses its mechanical properties and is increasingly subject to insect infestations that are difficult to see from the outside (Table 2.3) [12].

2.3 The Bamboo's Anatomical and Chemical Structure

Although the principal constituent components of bamboo are lignin, cellulose and hemicellulose—similar to wood—the microscopic structure of the two materials has

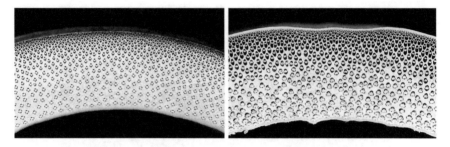

Fig. 2.9 Transversal sections of the walls of a *Moso bamboo* culm (left) and *Guadua angustifolia* (right) as seen through a microscope. Reproduced from De Vos [14]

differences, in particular in the layout and structure of the fibres, which contribute to different mechanical behaviour.

By dissecting the culm wall to the height of an internode, it can be seen that the lymphatic vessels create a progressively thicker pattern, from the inside to the outside of the section. The walls of these vessels are made from cellulose fibres [13]. The matrix in which the fibres are immersed is called parenchyma; these cells are longitudinally elongated and lignify from the initial phases of the plant's growth.

The microscopic structure of bamboo, therefore, makes it a natural composite material, in which the increased concentration of fibres towards the outside of the culm wall gives it a significant flexural rigidity (Fig. 2.9) [14].

Fibres make up 40–50% of the tissues and 60–70% of the culm's weight. They are longer than wood, but, more importantly, they have a higher tensile strength, as evidenced by their frequent use in the textile and engineering sectors. This is due to the walls of the fibre being stratified in concentric layers: the thicker leaves are oriented approximately along the axial direction, while, alternatively to these, thin layers of leaves are arranged transversally at various angles. This stratification is not found in wooden fibres (Fig. 2.10) [15].

The composition of the fibres depends on the species, the growth conditions of the plant and the age of the bamboo. As previously mentioned, the chief components are:

- cellulose and hemicellulose (together, these comprise more than 50% of the tissues);
- lignin (approximately 25%);
- pentosans (carbohydrates that account for 20–30%);
- resins, waxes, starches, tannins and inorganic salts that are found in smaller quantities;
- silica, the average content of which is between 0.5 and 4%.

This is particularly concentrated in the external surface layer of the culm, which is only one millimetre thick. This ensures that the element is naturally protected against external agents and gives the material exceptional resistance to fire. In fact,

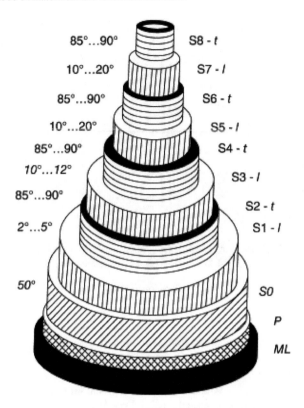

Fig. 2.10 Stratification of the wall of a bamboo fibre. Model of the structure of the fibre wall of bamboo. ML = middle leaf, P = primary wall, S = secondary wall, l and t indicate the longitudinal and transversal orientation of the microfibrils, respectively. Reproduced from Liese [15]

the ignition periods are greater than those of wood and the nodes act as a fire barrier against the spread of flames along the element.

The bamboo culm can be considered as a hollow tubular profile of natural origin from the construction point of view. The culm is the aerial segment that develops starting from the rhizome and is comprised of the neck, nodes and internodes (Fig. 2.11). The rhizome is typically an underground culm that comprises the support structure for the plant.

Nodes are the transversal septa that divide the internal cavity of the tube into closed chambers; when the culm is exposed to the heat of the fire, the air found inside the hermetic internodes expands to create a loud explosion, the sound of which probably gave rise to the onomatopoeic word "bamboo". The internode is generally hollow and frequently has a white exudate on the surface; however, in certain species, such as *Guadua amplexifolia* or *Guadua trinii*, the thickness of the culm wall completely saturates the internal cavity.

The nodes can be very prominent, such as in *Guadua paniculata* and *Guadua sarcocarpa*, or barely noticeable. The distance between these varies from one species

Fig. 2.11 Left: the neck of the culm. Right: a longitudinal section of an internode. Reproduced from Dunkelberg [1]

to another and is between one and four diameters, in proportion to the height of the plant. In general, there is also a progressive increase in the distance between the diaphragms from the base towards the central zone of the culm and, therefore, a reduction towards the apex, where the culm loses its axiality (Fig. 2.12).

The diameter of the culm is already defined starting from the early growth phases of the plant and the typical height of each species is reached within the first year. In the

Fig. 2.12 Types of bamboo culms. Reproduced from Van der Lugt et al. [7]

years that follow, the strengthening and thickening of the culm wall is accompanied by an increase in lignin and cellulose in the vegetative tissues.

From a structural perspective, the main reference parameters in terms of size are the diameter and thickness of the culm walls.

When the climatic conditions are favourable, certain species can reach 30 cm in diameter and have walls that are 30 mm thick: this is true of *Dendrocalamus giganteus* or *Guadua angustifolia*, which are capable of reaching heights of 35–40 m and grow up to 120 cm per day.

2.4 Durability of the Material and Protective Treatments

The main concern regarding the use of bamboo is undoubtedly its durability. The durability of bamboo elements, which are used without treatments, is inferior to the majority of woods and is largely dependent on the species, the section of the culm being considered, the thickness of the walls and the age of the plant when it was cut [16]. Information about the degree of durability for various species of bamboo is still scarce and further research would be needed to identify which varieties are more durable.

Table 2.4 gives a rough idea about the durability of untreated bamboo, based on various usage conditions.

One of the characteristics that negatively impacts the durability of the culms is their physical conformation, which includes thin walls and internal cavities in which insects can live. An attack from xylophagous insects—which could only impact the surface of a wooden beam, leaving the resistant section unchanged—is a much more serious risk for bamboo culms, since the thickness of the walls is no greater than a few centimetres.

The lower portion of the culm is considered to be more resistant to deterioration than the rest, while the internal part breaks down more rapidly compared to the outer section, making it difficult to evaluation the damage during the initial phases of the attack; this is due to the anatomical and chemical structure of the tissues, since only the outer surface layer is rich in silica. Other substances, such as waxes, resins and tannins, are found in modest quantities, so they do not improve the durability of the culms.

The large quantities of starch found in the bamboo make it easy for xylophagous insects (termites and beetles) to attack, as well as moulds and fungi. Tests have shown

Table 2.4 Durability of bamboo [13]	Duration (years)	Usage conditions
	1–3	Outside and in contact with the soil
	4–6	Under a cover and not in contact with the soil
	10–15	Under optimal storage/usage conditions

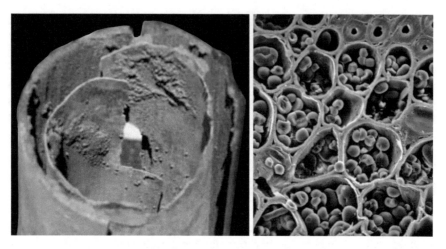

Fig. 2.13 Left: effects of the attack of xylophagous insects on a Gigantochloa sample that has not been treated and has remained in contact with the ground for a period of 8 months. Reproduced from Schröder [11]. Right: starch granules found in parenchyma cells. Reproduced from Liese [15]

that bamboo is more often subjected to soft decay and white decay than brown decay (Fig. 2.13) [17].

Although not all species are equally vulnerable to attacks from pathogenic agents, we must remember that the age of the plant when it was cut, any treatments and the way it was dried are all factors that influence its resistance to attacks.

The research carried out by Walter Liese shows that species that are more likely to encounter pathogenic attacks have high levels of starch in their culms. Nutrients (such as starch) are accumulated by the plant during the dry season and are then progressively consumed during the rainy season, as a result of the advent and development of new shoots. For this reason, we cut the culms (above the first above-ground node) at the end of the rainy season and, possibly, before dawn, when the plant's metabolism is at its lowest and the present of starch in the tissues is minimal.

Before describing the various types of treatment, we must supply some general instructions about storage and transportation, which are useful for preventing the premature destruction of the material.

During the storage phase—both before and after treatments—the culms must be kept dry, in a place that is protected from rain and appropriately ventilated; they must be away from the ground and stored in overlapping horizontal layers, ensuring the circulation of air between one element and another.

The transportation phases are also a critical moment for the durability of the material: the climatic conditions within a container, during sea transport, for example, are ideal for most fungi and insects. The treatment, therefore, must be performed before transportation and closed containers must not be used for storage.

It is essential to highlight that—beyond specific treatments—the most effective way to limit the depreciation of the material is to tackle the problem during the design phase, protecting the culms from water and light as much as possible with appropriate covers and from ground-based issues. No chemical treatment can be effective enough to correct a flawed design for the technical elements.

Most treatments that are suitable for wood are not appropriate for culms, due to the differing physical and botanical features of the bamboo. As previously mentioned, the external surface layer of the culm contains a high percentage of silica, which, despite protecting the culm from insects, forms a barrier against preservatives. Furthermore, the internal surface is covered with a waxy layer that is waterproof. In cases where a preservative is used, therefore, absorption can only effectively occur through the lymphatic vessels. These comprise only 10% of the transversal section and close within 24 h of cutting.

It is possible to apply two types of protective treatments, which are distinguished as physical (traditional) methods and chemical methods.

2.4.1 Traditional Methods

Traditional methods are aimed at reducing the amount of starch in the culms. Since these require no specific skills or financial resources, they are the most common methods.

- Treatment for transpiration: the cut culms are stored vertically, with their branches and leaves still intact. The transpiration process, which continues even after the plant is cut, causes the starch content in the culm to progressively decrease for a period of at least four weeks, after which the branches are removed.
- Immersion: immersion in water (or saltwater) for a period of 4–12 weeks leads to the removal of starches and sugars.
- This is one of the most commonly used methods, but it is not recommended as it may lead to the appearance of spots along the culms and cause the mechanical properties to weaken. It also does not ensure long-term protection from biological attacks on its own but improves the permeability of the bamboo for subsequent treatments (Fig. 2.14) [18].

  In certain cases, the culms are immersed in sand or mud and then dried in the shade in a ventilated environment. At this point, the maturing phase begins.

  During this time, the culms become more resistant and easier to work.

- Smoking: this involves exposing the culms to the smoke produced by burning waste bamboo or wood until they are saturated (three or four weeks). The process requires a temperature of approximately 50–60 °C and is particularly effective since the pores that are blocked by the combustion products prevent the entry of parasites. This last technique originated in Asia and significantly changes the natural colouring of the bamboo.

Fig. 2.14 Immersion treatment. Reproduced from Schröder [11]

2.4.2   Chemical Methods

Chemical preservation methods are more effective than traditional methods and are essential for large-scale projects, though they are generally more harmful to health and the environment. An aqueous solution with one of the following compounds is typically used as a preservative:

- sodium hydroxide;
- compounds of boric acid, borax and boron;
- sodium carbonate.

Sodium hydroxide treatments should not last for a long time, since this is an aggressive substance that can damage the material. Boric acid and borax, which are also used to treat wood, are not considered to be damaging to people and the environment and are authorised for use internationally. However, they may irritate the skin if inhaled or touched.

The most widely used treatment methods include boiling, immersion or injection (making small holes on the surface of the culm).

Some useful first steps that ensure better absorption of the preservative include the physical removal of the surface skill and the creation of holes in the diaphragms within the culm, which form a barrier to liquids entering.

Once the treatment phases, which may last from several hours to a few weeks depending on the type, the level of mechanisation involved in the process and the preservative used, are finished, the culms are left to mature.

2.4.3 *Drying*

Culms that have not matured cannot be used as building materials, due to the increased risk of biological attacks and the size changes caused by the progressive decrease in liquid content. This last factor determines a regular variation between highly heterogeneous values based on age and species in bamboo that has not yet been cut.

For example, when cutting 3–4 year old culms, Walter Liese's research shows that the relative moisture content varies between 100% at the bottom and 60% at the top and that, when the thickness of the culm wall is taken into account, the highest values are recorded towards the interior. Previous studies by the same author show that the amount of water retained in the tissues is closely linked to the amount of parenchyma present, but that the most important factor is the season, with variations of more than 50% moisture being recorded at various points during the year.

While maturing, the progressive loss of water—from its green state to 12–20%—causes the material to begin to shrink (mature culms are less prone to shrinkage than younger culms):

- maximum deformation occurs in the radial direction, with the thickness of the walls being reduced by 4–14% and the diameter of the culm reducing by 3–12%;
- minimal deformation occurs in the axial direction, to a sum of less than 1%;
- in the tangential direction, it is noted that the phenomenon is more significant on the external part of the culm than on the internal part.

The drying of the culms must take place in a dry and ventilated place, avoiding direct contact with the ground, for a period between 6 and 12 weeks. This can increase or decrease depending on the initial moisture content, the thickness of the culm walls and the thermo-hygrometric conditions of the storage location.

Drying is faster in a vertical position than a horizontal one, but sufficient supports are needed to avoid the flexural deformation of the culms (Fig. 2.15) [19].

From an economic point of view, it is estimated that, on average, the treatment and maturing of the culms incurs an increase of approximately 30% to the cost of the untreated raw material. In light of this limited increase in price, the useful life of a suitably treated culm increases by up to 15 years outdoors and up to 25 years if it is protected from rain and direct sun exposure.

2.5 Mechanical Properties of the Material

To this day, it is difficult to determine the mechanical performance of bamboo, since the resistance values are influenced by a myriad of factors, such as the moisture content, the species, the age of the bamboo when it is cut and the amount of fibres present. Even the same species may have different mechanical performances depending on altitude, climate and the type of earth in which the plant developed.

Fig. 2.15 Maturing bamboo culms, stored in a horizonal position. Reproduced from Schröder [11]

In Colombia, the best examples of *Guadua angustifolia*, which grow at altitudes between 900 and 1800 m, have notable greater performance than the same species in nearby Ecuador.

The most important of the parameters that must be taken into consideration is probably specific weight. This varies from 500 to 800–900 kg/m^3 and progressively increases towards the top of the bamboo, as does the fibre content. The external part of the culm wall has a specific weight that is significantly greater than the internal part, confirming the proportional relationship between weight and the amount of fibre present. The variation of the mechanical properties, therefore, more notable in the transversal direction than it is along the axis of the culm.

The mechanical resistance of the bamboo varies along the culm, so studies tend to distinguish three portions along the longitudinal axis:

- the upper portion, which has more fibres, is more resistant to compression and flexure;
- the central portion is more resistant to traction;
- the lower portion typically has worse levels of mechanical resistance.

Furthermore, it is necessary to identify different mechanical behaviour between nodes and internodes. The former are characterised by a reduction in the mechanical properties under any kind of stress, which is caused by the discontinuity in the section and the deviation of the fibre bundles; in the internodes, on the other hand, there is a greater level of resistance, specifically in the central portion, due to the presence of longer fibres that are aligned longitudinally on the axis of the culm [20].

Table 2.5 Mechanical performance of various species of bamboo originating in Asia [22]

Species	Origin	Specific weight (kg/m^3)	Moisture content (%)	Elastic module (N/mm^2)	Tensile strength (N/mm^2)	Compression strength (N/mm^2)
Bambusa bambos	India	650	15.5	6500	67.4	48.3
Bambusa vulgaris	Bangladesh	680	12.5	14,600	76.2	45.8
Bambusa natans	Bangladesh	680	12.8	12,900	87.7	71.8
Dendrocalamus strictus	India	720	10.7	15,949	118.4	64.5

Despite the ambiguity of the results—which often cannot be compared due to the countless factors that influence the mechanical properties of bamboo culms—it is possible to provide reference values as an example, such as the case of *Guadua angustifolia*, that afford a sufficient level of safety [21]:

- elastic module: 10,000–20,000 N/mm^2;
- resistance to compression: 40–60 N/mm^2;
- resistance to traction: 100–150 N/mm^2.

Guadua angustifolia is one of the species with the highest level of mechanical performance and is mostly used in Latin America.

A study published by Sattar in 1995 effectively summarised the results obtained by numerous studies about the mechanical performance of Asian bamboo [22].

For several years, studies have been carried out at the *Università Politecnica delle Marche* to evaluate the mechanical resistance of culms originating in Italy [23].

The species studies are *Phyllostachys edulis* (also known as *Moso*), which comes from the Italian Bamboo Centre in Genoa, and *Phyllostachys viridiglaucescens* from the bambusetum in Lucca and Selva di Paliano in Rome.

The average resistance values obtained are, on average, lower than those in the literature for the bamboo species in question, probably due to the high moisture content of the samples, which measures between 20 and 65%, and the fact that most of the samples came from the lower or central portion of the culm (Tables 2.5 and 2.6).

2.6 Structural Connection Technology for Bamboo Elements

The exceptional mechanical performance—which is noted through tests on individual bamboo elements—must be associated with the sufficient bearing capacity of the joints, which are often the weak points of the overall system. In fact, the connections between bamboo culms play a vital role in the structural capacity of the overall

Table 2.6 Mechanical resistance of culms originating in Italy [23]

	Phyllostachys edulis (Ge) (N/mm^2)	*Phyllostachys viridiglaucescens* (LU) (N/mm^2)	*Phyllostachys viridiglaucescens* (RM) (N/mm^2)
Compressive strength	56	57	69
Tensile strength	127	159	–
Flexural strength	97	–	–

construction, transferring the forces from one element to another and, therefore, to the ground. In other words, the connection system must ensure structural continuity, the safety transfer of loads and control of deformations.

In 1993, Arce-Villalobos identifies the requirements that a connection system for bamboo elements must adhere to [24]:

- *Maximising the use of the bamboo.* Like any building material, bamboo has positive and critical properties; the joint must make it possible to take full advantage of the material.
- *Simplicity.* Bamboo structures are primarily intended for areas where there is no advanced equipment or highly skilled labour. This is particularly important when working in contexts that lack resources, in self-construction projects where workers are often volunteers.
- *Stability.* The joint should ensure stability and durability over time, in relation to the expected useful life for the overall structure.
- *Adaptability to a modular production system.* Modularity makes it possible to focus technical issues in specific production areas, encouraging the participation of non-specialised workers on the construction site.
- *Predictability of mechanical resistance.* This theme is not sufficiently dealt with in literature and requires the study of virtual and mechanical models. As previously mentioned, the regulations in many countries where bamboo has been traditionally used as a building material, such as China and Colombia, provide models and reference standards.
- *Cost.* Joints often represent one of the most significant expenses when building a structure. It is not sufficient to compare the production cost of two different solutions, since the real economic impact of a connection system depends on factors like the construction time for the entire project, the necessary work, the amount of material used, the level of specialisation for the labour, the architectural possibilities, etc.

The creation of joints between bamboo elements is rather complex, thanks to certain characteristic features of the culms, which Arce-Villalobos defines as "internal constraints". This, however, is inevitable since we are talking about a natural material: the anisotropy of the mechanical behaviour and dimensional heterogeneity.

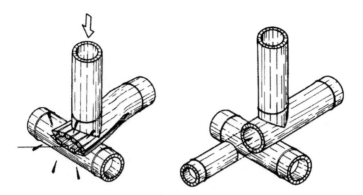

Fig. 2.16 Insertion of a stiffening cylinder to prevent the culm, which is subjected to loads that are transversal to its axes, from collapsing. Reproduced from López [25]

Bamboo is a highly anisotropic material, which shows strong mechanical properties when stress is placed in the longitudinal direction of the culm's axis, that is, parallel to the fibre. This is due to the lack of fibres in the radial or transversal direction and the weak resistance of the lignin, which the matrix found between the fibres is comprised of.

This property determines the tendency of the ends of the shaft to split when under compression or shear stress (Fig. 2.16) [25].

Common practice recommends placing transversal loads on the culm nodes, which are naturally made stiff by the presence of internal diaphragms. However, these are distributed along the axis of the shaft at variable intervals, making this solution restrictive from a construction point of view.

The tendency to split longitudinally is a well-known property to inhabitants of villages in areas where bamboo grows naturally. By placing a blade parallel to the fibres, they easily remove strips that they use to create baskets and other accessories (Fig. 2.17).

The imperfect circular nature of the section and the tapering of the culm section by height, accompanied by a progressive reduction in the thickness of the walls, are all factors that make the creation of connective systems even more complex. Given the heterogeneity of the sections, each connection represents a specific case, with problems that are always different and require work from highly specialised labourers. Despite the low cost of the raw material, the supplementary process that are necessary significantly increase the manufacturing costs for bamboo.

2.6.1 Classification of Bamboo Connections

Janssen's classification of the types of connections is based on three criteria [13]:

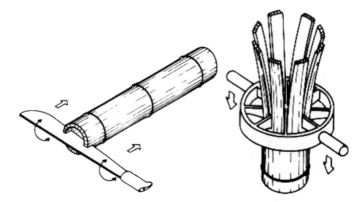

Fig. 2.17 Longitudinal cutting of the shaft to obtain strips, through the use of a blade or a "bamboo splitter". Reproduced from López [25]

- the connection can be made through direct contact between the elements or through a joint.
- the transfer of the force can happen by means of the internal or external surface of the culm, or through the entire section.
- the joint element can be parallel or perpendicular to the fibres.

By following these criteria, the Dutch architect defines six possible groups.

- *Joints that involve the entire section.* This group includes most of the traditional connection solutions, which use ties to keep the culms in the correct position (Fig. 2.18).
- *Joint that connects the internal surface of the culm to an element that is parallel to the longitudinal axis.* In this case, the cavity at the end of the culm is filled with mortar or a wooden element. The forces are transferred by adhesion from the internal wall of the culm to an element parallel to the longitudinal axis of the culm. This kind of configuration increases the resistant section and, as a result, improves the behaviour of the culm with respect to tangential stresses, which often determine the fracture of the fibres at the ends. If mortar is used, shrinking and the different hygroscopic behaviour of the two materials that are in contact can determine the loss of adherence between internal surfaces of the bamboo and said mortar (Fig. 2.19).

  In 1993, Arce-Villalobos proposes a connection made by gluing a wooden element inside the culm [24]. Two slits are formed at the end of the culm to control fractures when inserting the wooden cylinder (Fig. 2.20).

  The latter has a standardised size, while the culm will have a section that is an imperfect circle with a variable diameter. To this end, the internal surface of the culm must be smoothed, decreasing the thickness of the wall to a maximum of 5 mm, in order to obtain the desired diameter. The operation can be carried out using abrasive paper discs, which can be applied at the end of a drill. This is also useful for cleaning and preparing the internal surface for the subsequent gluing

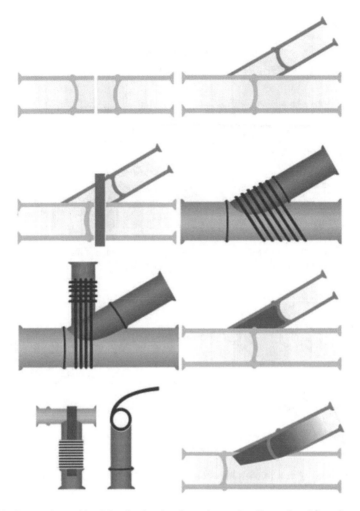

Fig. 2.18 Connections with a joint that involve the entire section. Reproduced from Janssen [13]

phase. It is then possible to use the same cylinder for culms with diameters that
vary by ± 1 cm compared to the wooden element.

- *Joint that connects the section of the culm to an element that is parallel to the
  longitudinal axis.* These systems require the presence of elements parallel to the
  longitudinal axis, which are fixed by means of transversal pins made from wood,
  bamboo or steel (Fig. 2.21).

Often, as happens in many structures by Simon Velez, including the Zeri
Pavilion for the 2000 Expo, it is possible to identify this system, but applied
through the use of mortar; the metal bar is anchored to a bolt by a hook, which
acts as a transversal pin and ensures the joint has greater resistance to traction.
The overall behaviour of the connection will be a mix between the second and

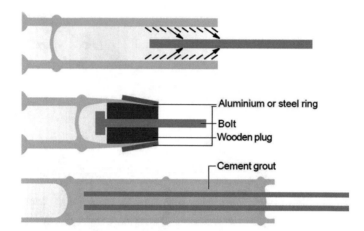

Fig. 2.19 Connections with a joint that connects the internal surface of the culm to an element that is parallel to the longitudinal axis. Reproduced from Janssen [13]

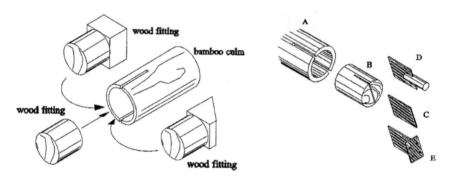

Fig. 2.20 Solution with a wooden insert inside the culm and a metal plate. Reproduced from Arce-Villalobos [24]

third group: part of the stress will be transferred to the culm by adhesion between the internal wall and the mortar, while the other portion will be transferred by means of a transversal bolt.

- *Joint that connects the section of the culm to an element that is perpendicular to the longitudinal axis of the culm.* The stresses are transferred from the section of the culm to a transversal element, such as, for example, a pin or a plate. This system can have risks, such as the bamboo breaking at the point where the pin is placed and a greater tendency towards insect attacks, which find shelter within the culm, since it is open at the ends (Fig. 2.22).

  The ITCR (Instituto Tecnológico de Costa Rica), for example, developed a connection system that uses wooden plates, which are inserted and glued to the slits that are made at the ends of the culms. Of course, this solution is only valid for rods that belong to the same plane [26].

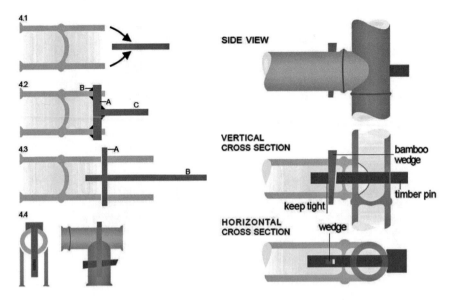

Fig. 2.21 Connections with a joint that connects the section of the culm to an element that is parallel to the longitudinal axis. Reproduced from Janssen [13]

- *Joint that connects the exterior of the culm to an element that is parallel to the longitudinal axis of the culm.* Traditionally, the connection through the external surface of the culm is made with ties, while the modern technique involves the use of metal straps (Fig. 2.23).

   In the case of ties, the connecting element comprises cords made from vegetate fibres. The transfer of the stresses is entrusted entirely to the forces of friction between the cords and the external surface of the culms (Fig. 2.24).

   This does not ensure that the joint is sufficiently resistant. With time, it inevitably deteriorates until the structure's mechanisms become unstable. However, thanks to how easy it is to find the necessary raw materials and the low costs, the tying system proved to be an efficient low-tech solution for rapidly building temporary structures after emergencies.

   The fibres are used when they are still green—or they are at least moistened—to ensure greater flexibility; as they dry, they shrink sufficiently to ensure that the connection is tightened.
- *Joint for split bamboo.* Connecting the bamboo strips is quite simple and can be done using metal plates, screws, bolts or glue.

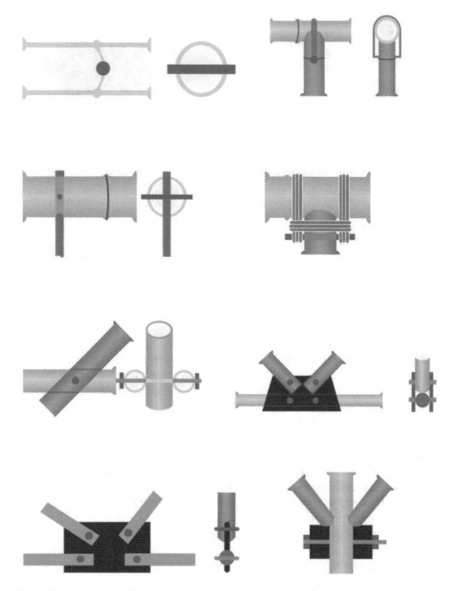

Fig. 2.22 Connections with a joint that connects the section of the culm to an element that is perpendicular to the longitudinal axis. Reproduced from Janssen [13]

2.6.2 Modern Experiments on Connections Joints

In the past thirty years, experimental studies have been carried out regarding the production of connection joints that facilitate efficient construction systems from a

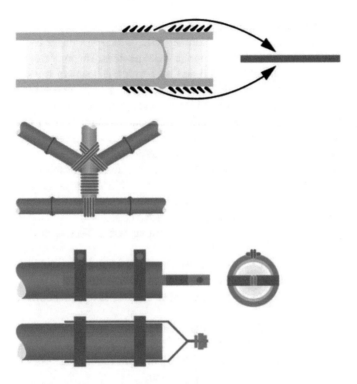

Fig. 2.23 Connections with a joint that connects the exterior of the culm to an element that is parallel to the longitudinal axis. Reproduced from Janssen [13]

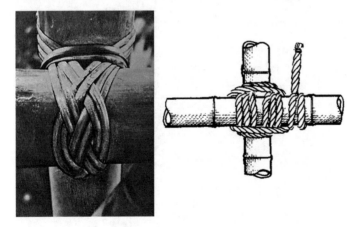

Fig. 2.24 Left: knotted binding. Reproduced from Dunkelberg [1]. Right: squared binding. Reproduced from López [25]

performance and implementation point of view. The technological solutions have indirect connections between the bamboo rods, which are made by inter-laying elements; most of the time, these comprise metal devices. The joint may take advantage of the hollow shape of the culm and affix to it by means of transversal bolts. Other solutions require the connection between the node and the culm to be ensured using a third hinge element. The rod is inserted and bolted into this, making it possible to design reticular structures with high performance qualities. The function of the metal elements plays a vital role in fixing the structure to the ground, since the steel plates can be easily fixed to the reinforced concrete base.

Based on the research carried out, it is clear that this type of mechanical joint—where the culm is not shaped—is more common in Western construction, since it is a priority to save on the use of specialised labour and reduce the high costs of supply raw materials imported from the East or Latin America.

Some specific technological solutions for indirect connections with different metal devices are listed below.

The Japanese architect Shoei Yoh developed a system for a roof in Fukuoka in 1989. This involved inserting steel tubes at the ends of the culms and fixing them using bolts. A plate welded to the tube makes it possible to connect the rods to the node (Fig. 2.25) [27].

In 1997, Renzo Piano proposed a connection system with welded plates in which the bolts are replaced by a binding wire tied around the bamboo. It was presented in the *Renzo Piano Building Workshop* (RPBW) in New York and allows a lightweight spatial structure to be produced [28] (Fig. 2.26).

Fig. 2.25 Connection proposed by the architect Shoei Yoh, Fukuoka, 1989. Reproduced from RWTH Aachen University

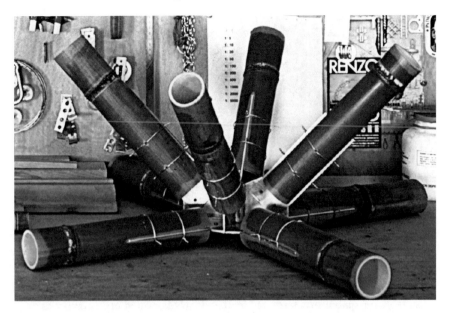

Fig. 2.26 Connection proposed by the architect Renzo Piano, New York, 1997. Reproduced from RWTH Aachen University

In 1998, the architects Cheyne and Londoño, in collaboration with the *Guadua Tech Team*, developed a connective system that was used to produce the *Bamboo Convention Center* in Colombia (2008). The assembly principle made it possible to standardise the sections of the culms. The ends had been carved prior, then tightened using a steel cable and fixed in a metal crown. A steel bar inserted inside made it possible to connect it to the node (Fig. 2.27) [29].

In 2006, the Italo-Colombian architect Mauricio Càrdenas Laverde developed, in collaboration with ARUP Italia, a dry joining system for the construction of a

Fig. 2.27 Guadua tech system

Fig. 2.28 Connection system developed by the architect Maurizio Càrdenas Laverde in collaboration with Arup for the production of the *Dagad* Pavilion (Milan)

bamboo and steel pavilion for the *Dagad* association at the Milan Fair. The system is not suitable for spatial structures, but merely flat roofs, which in this case was made from sheets of polycarbonate (Fig. 2.28) [30].

More innovative technological solutions, which combine bamboo, steel and fabric membranes, were designed by the artist Markus Heinsdorff, in order to construct structures for the "Germany-Chinese Esplanade-Moving Ahead Together" event, organised between 2007 and 2010 in 5 Chinese cities (Chongquing, Guangzhou, Shenyang and Wuhan, China). The type of bamboo used is *Phylostachys pubescens*, known as "Mao bamboos" or "moso" in China (Fig. 2.29) [31].

In an attempt to design new prototypes of sustainable homes for the island of Haiti in the aftermath of the 2010 earthquake, the architect Laurent Saint-Val proposed a project of vertical bamboo towers in 2011. These took their inspiration from the traditional art of wicker (weaving natural vegetate fibres). The connection system of bamboo culms with metal X-shaped joints formed the exo-skeleton of the structure. The technological solution for the assembly required the use of the patent conceived by the *Guada Tech Team* (Fig. 2.30) [32].

Also in Colombia, the military aviation in Cali required the construction of a hangar for vehicles that were no longer in use; a system not unlike the Mero[2] one was trialled. This was a solution for connecting reticular structures with tubular steel rods designed by Jörg Stamm. This system is very similar to the previous one, but it does not require the use of the steel cable that wraps around the end of the culm (Fig. 2.31) [33].

In recent years, numerous proposals about modern connective systems for bamboo structures have been proposed in universities and then trialled by students in educational workshops [34]. In the Faculty of Architecture at the University of Turin, for example, the architect Simone Cantoni has studied a steel joint.

---

[2] Mero construction system with spherical nodes and tubular rods.

Fig. 2.29 *Diamond Pavillion*, details of the joining system designed by Markus Heinsdorff for the "Germany and China—Moving Ahead Together" event series, China, 2007–2010. Reproduced from Minke [31]

Fig. 2.30 Design of the vertical towers by the architect Laurent Saint-Val, Haiti, 2011. Details of the joint

Fig. 2.31 Bamboo aircraft hangar. Mero-style connection system for the production of a hangar in Cali (Colombia)

The operating mechanism makes it possible to expand four plates inserted at the end of the culm, which put pressure on the internal surface. This is balanced externally by metal clamps. An additional circular plate, on the other hand, presses against the diaphragm of the node. This system has the benefit of being adaptable to culms with different diameters and is completely reversible: both the joint and the culm can be re-used (Fig. 2.32) [35].

In April 2019, on the occasion of the *International Horticultural Exhibition* in Yanquing (China), the INBAR Garden Pavilion, known as the "Bamboo Eye" Pavilion is inaugurated. This imposing building—the largest of its kind to ever be built in northern China—was designed by the previously mentioned Italo-Colombian architect Mauricio Cardenas. The intention of the design was to create a large, natural and sustainable open space that was covered by a trussed structure of over 1600 m^3 in size. This comprised large arches made from rounded and pre-curved bamboo culms (*Phyllostachys pubescens*, 8–10 cm in diameter and 4–6 years old), which extended for approximately 40 m with a slope of 9 m (Fig. 2.33) [36].

Fig. 2.32 Detail of the reversible joint designed by Cantoni. Reproduced from Cantoni [35]

Fig. 2.33 INBAR Garden Pavilion, Beijing International Horticultural Exhibition in Yanquing (China), designed by the Italian-Colombian architect Mauricio Cardenas

2.7 The Bamboo's Legislation and Market

Today, technical regulations tend to limit the use of bamboo as a building material, particularly in Europe. This is because there are not sufficient assurances regarding the homogeneity of the mechanical behaviour of the culms, even where these come from the same species.

The first step was taken in 2004, when the physical and mechanical properties of bamboo were certified with ISO/DIS 22156 *"Bamboo Structural Design"* and ISO/DIS 22157 *"Determination of physical and mechanical properties of bamboo"*. On the other hand, detailed technical regulations have been adopted in countries with a long history of using bamboo as a building material, such as India, Peru, China, Colombia and Ecuador.

The ISO/DIS 22156 and ISO/DIS 22157 certifications, which were introduced in 2004, are the result of a long period of work that was chiefly carried out by INBAR (*International Network for Bamboo and Rattan*) since 1997. This was done with the aim of having bamboo recognised as a building material on a global level. These highlight the importance of traditional bamboo construction techniques which, despite being widely disseminated and recognised, may constitute non-codified "standards" that can be applied in similar contexts.

According to the aforementioned regulations, a structure will be considered sufficiently reliable if there are reports containing descriptions of similar structures that have resisted natural phenomena, such as earthquakes and hurricanes. These reports must be drafted by engineers with suitable sector experience.

The ISO/DIS 22156 standard says that characteristic resistance may be obtained as expressed below:

$$R_k = R_{0.05} \cdot \left(1 - \frac{2.7 \frac{s}{m}}{\sqrt{n}} \right)$$

In which:

R_k = Characteristic strength,

$R_{0.05}$ = Resistance of the 5th percentile,
m = average value obtained from the tests,
s = standard deviation,
n = number of tests carried out (at least 10).

It is assumed that the material is elastic and, therefore, that it has a linear relationship between force and deformation until it breaks. The plastic deformation section is to be considered as insignificant.

The loads and actions that must be considered in the calculations are those that are defined by the national regulations for the type of structure in question.

Instead of the procedure used for the ultimate limit state, the project can be carried out by considering the maximum allowable stress, which can be calculated in the following manner:

$$\sigma_{all} = \frac{R_k \cdot G \cdot D}{S}$$

In which:

σ_{all} = allowable stress in N/mm^2,
R_k = characteristic strength,
$G = 0.5$ (factor that considers the difference between the results obtained from the lab tests and those expected in the field),
$D = 1-1.5$ (coefficient that depends on the duration of the load),
$S = 2.25$ (safety factor).

For permanent loads, with a standard deviation of 15%, the allowable stress will be 1/7 of the average value ("m") of resistance obtained from the tests.

China is currently the largest exporter of bamboo as an unprocessed raw material, obtaining a turnover of $48.2 million in 2012, which is approximately 70% of the global market. This is followed by Europe with $10.9 million and then Colombia, Pakistan, Vietnam and Indonesia, which have increasingly smaller market shares [37].

On the other hand, Europe, followed by the United States and Japan, is the largest importer of the raw material, with a value of $53.4 million. This comprises 50% of the imports of bamboo globally. If we analyse data related to industrialised products derived from bamboo, rather than its supply as a raw material, such as plywood panels, laminate flooring and items made from bamboo paper, the global dynamic between importing and exporting countries is, on the whole, unchanged: China is the largest producer and exporter with $413 million turnover in 2012 and 77% of the market share. Meanwhile, Europe imports nearly half of the products deriving from bamboo (44%), followed by the United States (11%) and Canada.

In terms of the market within the European Union, it is possible to say that Italy, along with Belgium and Spain, and after the Netherlands, is one of the greatest exporters of products made from bamboo. However, the data collected in this field is not strictly tied to the construction sector, but also includes the textile and furnishing markets.

2.8 The Bamboo Supply of the Botanical Garden of Palermo

For the purposes of this research, the Director of the Botanical Garden of Palermo authorised the supply of approximately 25 linear metres of bamboo from the *Bambusa vulgaris* and *Phyllostachys viridis* species. The available culms do not have characteristics that are suitable for structural purposes, due to both the species and the age of the plants, but they have made it possible to perform all of the necessary tests, thus developing a method that can also be applied in similar contexts (Fig. 2.34).

Data regarding the dimensional properties was collected for the bamboo, with each culm then ideally sub-divided into three segments: bottom (B), middle (M) and top (T). This sub-division is generally adopted to consider the different properties of the culm in its longitudinal development. Since the section of the culms is oval-shaped, the diameter is assessed as the average between the minimum diameter (D_1) and the maximum diameter (D_2). These measurements were taken at the lower and upper ends of each section (Fig. 2.35).

The culms were first weighed when they were collected, finding the weights P_1, and, subsequently, after a maturing phase of one month in a dry and ventilated environment, finding the weights P_2 (Tables 2.7 and 2.8).

This made it possible to calculate the specific weights γ_1 and γ_2, and the percentage difference:

$$\Delta\gamma = \left(1 - \frac{\gamma_2}{\gamma_1}\right) \times 100$$

Fig. 2.34 Supply of bamboo culms at the Botanical Gardens of Palermo

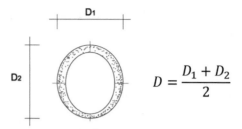

Fig. 2.35 Diameter of a generic section of a culm. Reproduced from Fabiani [23]

Table 2.7 Characteristics of culms from *Phyllostachys viridis*

Section	L	D_1	D_2	D	t	V	P_1	γ_1	P_2	γ_2	$\Delta\gamma$	r
	mm	mm	mm	mm	mm	mm^3	kg	kg/m^3	kg	kg/m^3	%	mm/m
1-B	1420	43	49	46	9	2,385,646	1.20	503	1.05	440	12.5	−0.4
		45	48	46.5	4							
1-M	1325	45	48	46.5	4	1,821,749	0.83	456	0.72	395	13.3	7.2
		36	38	37	4							
2-B	1420	42	48	45	9	2,183,968	1.22	559	1.12	513	8.2	1.1
		42	45	43.5	4							
2-M	1430	42	45	43.5	4	1,801,577	0.90	500	0.79	439	12.2	4.9
		36	37	36.5	3.5							
3-B	1340	48	49	48.5	5.5	2,301,234	1.11	482	0.97	422	12.6	2.6
		44	46	45	5							
3-M	1510	44	46	45	5	2,482,492	1.00	403	0.88	354	12.0	−1.0
		46	47	46.5	4							
3-T	1540	46	47	46.5	4	1,998,418	0.81	405	0.68	340	16.0	7.8
		34	35	34.5	3.5							

Section 1-B indicates that it is referring to the lower portion of culm 1; L indicates length, D is diameter, t is thickness, V is volume, P is weight, γ is specific weight, r is tapering of the culm

From the difference between the diameters at the end of each segment, divided by the length of said segment, we obtain a factor r that indicates the tapering, that is, the reduction of the diameter in millimetres per liner metre:

$$r = \frac{D_{inf} - D_{sup}}{L} \cdot 1000$$

Although it is not a statistically significant sample, it is possible to draw some conclusions from the data collected.

The specific weights are considerably lower than what is indicated in the literature, ranging from 500 to 900 kg/m^3. In the case of *P. viridis*, after the maturing period, γ

Table 2.8 Characteristics of culms from *Bambusa vulgaris*

Section	L	D_1	D_2	D	t	V	P_1	γ_1	P_2	γ_2	$\Delta\gamma$	r
	mm	mm	mm	mm	mm	mm³	kg	kg/m³	kg	kg/m³	%	mm/m
1-B	1565	58	60	59	15	4,242,507	1.45	342	1.40	330	3.4	0.3
		56	61	58.5	9							
1-M	1590	56	61	58.5	9	4,310,279	0.94	218	0.92	213	2.1	−0.3
		59	59	59	5							
1-T	1510	59	59	59	5	3,755,413	0.58	154	0.57	152	1.7	3.6
		52	55	53.5	4							
2-B	1310	56	60	58	9	3,461,124	0.84	243	–	–	–	0.0
		57	59	58	6							
2-M	1305	57	59	58	6	3,388,809	0.68	201	0.67	198	1.5	0.8
		55	59	57	5							
2-T	1420	55	59	57	5	3,024,971	0.58	192	0.57	188	1.7	7.0
		46	48	47	4							
3-B	1290	65	68	66.5	18	3,810,255	1.70	446	–	–	–	8.1
		56	56	56	11							
3-M₁	1560	56	56	56	11	4,159,319	1.65	397	1.60	385	3.0	−2.90
		59	62	60.5	7							
3-M₂	1710	59	62	60.5	7	4,636,926	1.20	259	1.18	254	1.7	2.0
		56	58	57	5							
3-T	1730	56	58	57	5	3,719,661	0.92	247	0.91	245	1.1	5.5
		47	48	47.5	4							

Section 1-*B* indicates that it is referring to the lower portion of culm 1; *L* indicates length, *D* is diameter, *t* is thickness, *V* is volume, *P* is weight, *γ* is specific weight, *r* is tapering of the culm

values of between 340 and 513 kg/m³ were found, while, for *B. vulgaris*, the values were between 152 and 385 kg/m³, with higher values are the base and lower ones near the top due to the decreasing thickness of the culm walls. This discrepancy is probably due to the age of the culms, which were not yet mature when they were cut.

Another aspect to note concerns the $\Delta\gamma$, that is, the percentage reduction in specific weight which, during the maturing period, averaged 12.4% for *P. viridis* and a mere 2% for *B. vulgaris*. This is due to the fact that the *B. vulgaris* culms were cut prior to being sampled.

In terms of the tapering of the culms along the longitudinal axis, it is noted that, on average, the diameter varies by 3.5 mm/m for *P. viridis* and 3.1 *mm*/m for *B. vulgaris*. In percentage terms, this means that a variation in diameter of around 8.2% and 5.4% respectively per linear metre has been recorded.[3]

---

[3] These values are in line with other studies, such as that of Fabiani Marco, who, in *Bamboo structures: Italian culms as likely resource for green buildings* (pp. 23–24), notes the maximum diameter-variation values per linear metre as 8.6% for *P. edulis* and 10% for *P. viridiglaucescens*.

We also wanted to determine the variation in the thickness of the culm walls per linear metre:

$$\Delta t = \frac{t_{inf} - t_{sup}}{L}$$

There was an average variation in thickness of 0.43 mm/m for *P. viridis* and 1.67 *mm*/m for *B. vulgaris*, eliminating extreme values. For percentage values, the ratio

$$\Delta t(\%) = \frac{\Delta_t}{t_{inf}}$$

was calculated, which was 9.2% and 18.8% per linear metre respectively.

It should be noted that the tapering factor, which is indicated with *r*, sometimes presents negative values; this stems from the fact that the area of the culm section experiences an increase in the lower section, proceeding from the bottom to the top. The section begins to shrink in the middle section, then we move to modest positive *r* values, while, in the apical section, larger positive values indicate a major reduction of the section per linear metre.

References

1. Dunkelberg K (1985) Bamboo as a building material. IL 31 Bambus, Institute of Lightweight Structures, University of Stuttgart
2. Laverde Càrdenas M (2005) Bambù, un materiale da riscoprire. Costruire 260:57–64
3. Ramponi R (2008) Acciaio Vegetale. Casa & Clima 14:56–61
4. Bar L (2012) Il mondo del bambù. Un futuro possibile e sostenibile. In: Bambù per ideare sperimentare e costruire. Aracne, Roma, p 16
5. Van der Lugt P (2005) Bamboo as a building material alternative for Western Europe? A study of the environmental performance, costs and bottlenecks of the use of bamboo (products) in Western Europe. Bamboo Rattan J 2(3):205–223
6. Janssen J (1981) Bamboo in building structures. PhD dissertation, Eindhoven University
7. Van der Lugt P, Van den Dobbelsteen AAJF, Janssen JJA (2006) An environmental, economic and practical assessment of bamboo as a building material for supporting structures. J Constr Build Mater 20:648–656. https://doi.org/10.1016/j.conbuildmat.2005.02.023
8. Firrone T (2006) Il bambù. Aracne, Roma
9. Conti ML (2006) Bambù: Botanica, Design e Architettura. Nuova Ipsia Editore, Palermo
10. Douglas S (1901) In Sicily V1: 1896–1898–1900. Sands and Co., London
11. Schröder S. Guadua angustifolia. https://www.guaduabamboo.com/blog/guadua-angustifolia. Accessed 21 Feb 2021
12. Cassandra Adams. Bamboo architecture and construction with Oscar Hidalgo. http://www.net workearth.org/naturalbuilding/bamboo.html. Accessed 20 Mar 2019
13. Janssen J (2000) Designing and building with bamboo. INBAR, Technical Report 20
14. De Vos V (2010) Bamboo for exterior joinery. PhD dissertation, Larenstein University
15. Liese W (1998) Anatomy and properties of bamboo. INBAR, Technical Report 18
16. Schröder S. Durability of bamboo. http://www.guaduabamboo.com/blog/durability-of-bamboo. Accessed 22 Mar 2019

17. Schröder S. Bamboo insect infestation. https://www.guaduabamboo.com/blog/bamboo-insect-infestation. Accessed 21 Feb 2021
18. Schröder S. Leaching bamboo. https://www.guaduabamboo.com/blog/leaching-bamboo. Accessed 21 Feb 2021
19. Schröder S. Drying bamboo poles. https://www.guaduabamboo.com/blog/drying-bamboo-poles. Accessed 21 Feb 2021
20. López OH (2003) Bamboo: the gift of the gods. Bogotà
21. Mechanical properties of bamboo. RWTH Aachen University. http://bambus.rwth-aachen.de/eng/reports/mechanical_properties/referat2.html. Accessed 4 Gen 2015
22. Sattar MA (1995) Traditional bamboo housing in Asia: present status and future prospects. INBAR, Technical Report 8
23. Fabiani Marco (2014) Bamboo structures: Italian culms as likely resource for green buildings. PhD dissertation, Università Politecnica delle Marche
24. Arce-Villalobos OA (1993) Fundamentals of the design of bamboo structure. PhD dissertation, Universiteit Eindhoven
25. López OH (1981) Manual de Construcción con Bambú. Bogotá
26. Jayanetti L, Follett P (1998) Bamboo in construction: an introduction. INBAR, Technical Report 16
27. Collettivo Cerretini (2010) Bamboo: Architettura, Tecnologia, Design. https://issuu.com/collettivocerretini/docs/bamboo_-_collettivo_cerretini/25. Accessed 3 Mar 2021
28. AV MONOGRAFÍAS Nº 119, Renzo Piano, Building Workshop (RPBW), Arquitectura Viva 1990–2006
29. Guadua Tech System. http://www.bamboocraft.net/gallery/showphoto.php?photo=1220. Accessed 5 Mar 2021
30. Laverde Càrdenas M (2006) Microclimati Pavilion. https://www.studiocardenas.it/index.php/it/28-projects/93-microclimatic-pavilion. Accessed 6 Mar 2021
31. Minke G (2012) Building with bamboo: design and technology of a sustainable architecture. Birkhauser Verlag Ag, Switzerland
32. Saint Val L (2011). Bamboo housing for Haiti. http://urbanlabglobalcities.blogspot.com/2011/03/bamboo-housing-for-haiti-by-laurent.html. Accessed 15 Feb 2021
33. Mero Construction system. http://www.meroitaliana.it/it_IT/prodotti/2/sistemicostruttivi. Accessed 06 Apr 2019
34. Bar L (2021) Bambù Strutturale. http://bambustrutturale.it/. Accessed 1 Mar 2021
35. Cantoni Simone (2007) Il bambù nelle costruzioni: Studio di un giunto reversibile per strutture reticolari in bambù. Thesis, Politecnico di Torino
36. INBAR Garden Pavilion—2019 Beijing International Horticultural Exhibition. https://www.studiocardenas.it/index.php/zh-cn/innovative-technologies/154-inbar. Accessed 1 May 2019
37. International trade of bamboo and rattan 2012. http://www.inbar.int/publications/?did=292. Accessed 1 May 2019

Chapter 3
Algorithmic Modelling and Prototyping of a Connection Joint for Reticular Space Structures

Abstract The detailed research process proposes innovative and advanced solutions that allow for the determination and parametric control of any configuration of the reticular spatial structure. The system allows a variable number of culms of the same size can be joined and oriented in generic directions, ensuring the designer has optimal freedom in terms of composition. The algorithmic definitions, which are processed in the *Grasshopper* plug-in for the famous NURBS modelling software *Rhinoceros*, have allowed the project flow to be managed as an integrated process, ranging seamlessly from the concept to the manufacturing with numerically controlled machines. The structuring of a design process in algorithmic terms has had the advantage of obtaining flexible solutions to the different design conditions through form-finding methods, which can be optimised with genetic algorithms. The full-scale, three-dimensional printing—using professional machines for rapid additive prototyping—of a spatial-reticular-structure module with various types of connection joints and rods in bamboo culms have permitted experimentation and validation of the defined technological solution.

3.1 Design of a Parametric Connection Joint for Sustainable Structures in Bamboo

As outlined in the previous chapter, the features of bamboo also make it an ideal material for use in architectural endeavours, even from a structural point of view. In terms of mechanical performance, it is certainly comparable—if not superior—to wood, while its high degree of renewability compared to the latter and its low cost as a raw material make it a valid alternative for any structures aiming at environmental sustainability.

The reasons why its use as a building material has been limited to date, especially in Europe, has also been analysed. Among these, the aspects that most deter architects and engineers from using bamboo are the lack of joining solutions that ensure appropriate levels of performance and the absence of national legislation on this topic.

The elevated mechanical properties of bamboo culms, in fact, rarely corresponds to an adequate structural capacity in the connection joints.

In order for bamboo to become a commonly used building material, it is necessary to develop appropriate construction techniques; this is a long process (it has required centuries of experimentation in the case of connections for wooden structures) that has not yet produced consolidated techniques, since the use of bamboo has thus far been limited solely to resource-free contexts.

When designing a new connective system, two problematic elements of the features of bamboo must be taken into account:

- the heterogeneity of the size of the culm sections;
- the anisotropy of its mechanical behaviour, which makes the ends of the culms prone to splitting along the longitudinal axis.

Our objectives were to create a solution that ensures:

1. the dry assembly of the structure, in order to simplify assembly and disassembly operations and potentially allow for the replacement of individual parts of the structure;
2. dimensional adaptability to a specific range of section diameters;
3. adaptability to a number of different rods that converge at the node;
4. adaptability of the rods that converge at the node to various generic directions, on order to afford the designer optimal formal freedom;
5. Formal continuity between rods and connection joints, such that the latter is not seen as an interruption, but as an integrative element between the parts from both a structural and aesthetic point of view.

Therefore, the design of the connection joint is intended to be as generic as possible and adaptable to any spatial configuration of the lattice of the rods that comprise the structure.

In order for the adopted solution to respond effectively to all of the requirements outlined above in a generic—or, rather, a parametric—manner, it was necessary to structure the problem in algorithmic terms.

3.2   Machining of the Culms' Ends

As has already been mentioned, one of the typical problems of connections for bamboo structures is adapting a specific range of section diameters to a solution that can be, at least in part, standardised.

The longitudinal fissure prevents the risk of splitting at the end of the culm when the wooden cylinder is being inserted and allows for the flexibility of the still-green bamboo to be exploited, in order for the section to be adapted to the cylinder, thus enlarging or restricting the two strips that have been obtained in this way.

The degree of flexibility will be proportional to the length of the fissure, which, in any case, will not exceed the length of the cylinder, given that it acts, to a certain extent, as a stopper, which prevents the entry of insects into the bamboo that would cause rapid deterioration.

It is legitimate to ask about the degree of adaptability, that is, the dimensional range, that this solution must be able to respond to. From the dimensional analysis of the culms taken from the Botanical Garden of Palermo, it has been noted that there is a tapering of the external diameter of the culms. This is, on average, between 5 and 8% per linear metre, depending on the species and the section under consideration (lower, middle or upper). In the case of a rod that is 2 m long and that has a diameter of 10 cm at one end, it will be possible to expect a diameter between 8.4 and 9 cm at the opposite end. In addition to the tapering of the section, there is also a reduction in the thickness of the wall, which has been assessed between 9 and 18% per linear metre. In fact, then, the internal diameter experiences smaller dimensional variations compared to the external diameter. Assuming that our rod with a 10 cm diameter has walls that are 1 cm thick, these will be reduced to 0.6–0.8 cm at the opposite end. In our case, therefore, the internal diameter will vary by 6–10 mm, based on the appropriate calculations. This is the parameter that must be considered in greater detail, since the system adopted must ensure a level of adaptability and a range of variability of the diameters that is of the same magnitude.

Based on the tests carried out, as well as comparisons with other studies, this solution allows, for a single diameter of the wooden cylinder, for adaptation to a variety of bamboo sections within a range of $\pm$ 1 cm compared to the cylinder, which is fully compliant with what the company ensures for its furniture [1].

For example, in the case of culms with a 10 cm diameter, the company generally ensures that these have a diameter between 9 and 11 cm (Figs. 3.1 and 3.2).

Furthermore, the use of an abrasive roller for drills allows the thickness of the walls to be easily reduced to 5 mm, thus allowing the cylinder to be adapted to a greater variety of diameters. In any case, this operation is useful for preparing the internal surface of the culm for potential bonding with epoxy resin or wood adhesive.

The chief advantage of the movement from a hollow section to a solid section is that this considerably increases the mechanical resistance of the end of the culm for the greater area and the greater moment of inertia.

$$A_0 = \pi \cdot \frac{(D^2 - d^2)}{4} \rightarrow A_f = \pi \cdot \frac{D^2}{4}$$

$$I_0 = \pi \cdot \frac{(D^4 - d^4)}{64} \rightarrow I_f = \pi \cdot \frac{D^4}{64}$$

where D is the external diameter of the culm, while d is the internal diameter.

In terms of the cylindrical wooden inserts, these must have a diameter that is approximately equivalent to the internal diameter of the culms that will be used to build the structures. The difference between the two sizes must be less than a centimetre:

Fig. 3.1 Longitudinal sectioning of the ends of the culms in order to accommodate a cylindrical wooden insert (top right) and production of the cylindrical wooden inserts (top left). Assembly of the culm and the wooden insert with bolts and a metal clamp (bottom right)

$$\left| d_{culm} - d_{cylinder} \right| \leq 1\text{cm}$$

The cylinders will have a length of 15 cm and a fissure of 11.5 cm, in order to allow the plate to be housed. These measurements should be considered as the minimum for ensuring an appropriate spacing of the bolts at the end of the bamboo.

Janssen's research shows that greater distances from the end offer greater structural performance for both short- and long-term loads, thus preventing the risk of the bamboo collapsing prematurely. Taking this research into consideration, it was decided that the bolts would be positioned at a distance of 5–10 cm from the end. The culm and wooden insert, the longitudinal fissures of which have already been produced, are solidified at this point by means of a metal clamp and drilled with a common wood tip [2].

It is recommended that the process described up to this point be carried out while the culms are still green, to ensure greater flexibility and adaptability of the strips.

During the subsequent drying phase for the material, which can last up to 4 months, the metal clamp is kept very tight, to ensure that the fibres assume the desired configuration during this time, progressively deforming through the *creep* phenomenon.

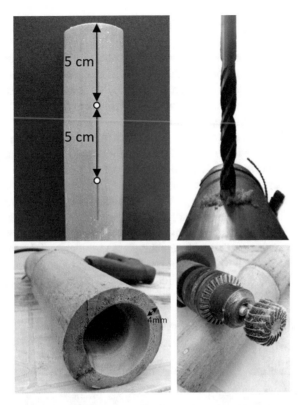

Fig. 3.2 Positioning of the holes (top left) and drilling of the culm (top right). Adaptation of the diameter thanks to the reduction of wall thickness with an abrasive shaver (bottom)

3.3 The Connection Plate

The shifting of force from the road to the connection joint occurs by means of a 5 mm-thick steel plate, which is inserted inside the wooden cylinder. The holes are oval-shaped, which allows the plate to be adjusted and perfectly aligned with the longitudinal axis of the culm. The end of the plate will be outside of the culm and has a round shape that ensures greater operational safety during the handling and installation phases.

In addition, the shape complies with the stresses that develop through an axial force applied to the end, in accordance with the bolt that connects the plate and the connection joint.

The following diagram, taken from *Millipede*[1] allows for the visualisation of the areas under greatest stress. The red points around the oval-shaped holes indicate the

---

[1] *Millipede* is a structural analysis and optimisation software that is distributed as a Grasshopper plug-in. Further information available at http://www.sawapan.eu/.

restraints, while the blue lines around the bolt indicate the area in which the force has been applied.

Starting from a rectangular plate, the topological-optimisation process has allowed the end of the plate to be shaped in such a way as to eliminate the areas that are under the least pressure. The plate has been parametrically designed so that it is possible to vary the dimensions of the plate and the dimension and position of the holes by simply working on a set of input parameters.

The design of the plate and the design of the connection joint are part of a single process: the parameters that control the thickness of the plate and the position and dimensions of the holes allow the plate and the connection joint to be worked on at the same time (Figs. 3.3 and 3.4).

If it is necessary to use a larger plate than expected, even at an advanced stage of the project, it will be sufficient to vary a parameter to update the plates and all of the connection joints in the structure, which will need to house larger dimensions.

At this point, it is possible to insert the plate at the end of the bamboo culms. The process described up to this point can be carried out away from the work site: all of the rods must be cut to the desired length and assembled, so that, during the

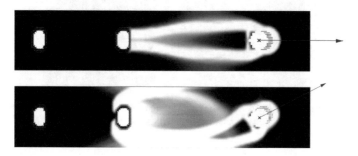

Fig. 3.3 The areas of the plate that are under most stress when a fictitious force is applied

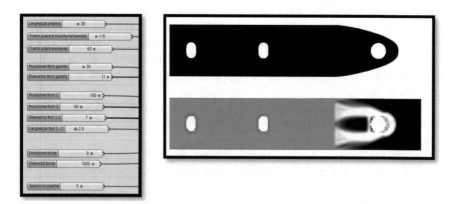

Fig. 3.4 Parametric modelling of the connection plate

on-site assembly and disassembly phases, the only operation that must be carried out is fixing the rod-connection joint system with a bolt.

During this experimental phase, in order to validate what has been exposed to this point, a prototype of a spatial-reticular structure is produced. This predicts the manufacturing, in 3D print, of the principal components (the plates and connection joints) and the assembly with the culms and wooden inserts.

For a description of the digital-manufacturing process, please refer to Chap. 3.4.

3.4 Algorithmic Connection Joint Modelling

From the geometric point of view, the connection joint must satisfy the adaptability requirement for any number of rods converging at the node, no matter what their orientation. Although it has been designed with the intention of solving the connection problem for bamboo structures, the proposed connection joint is also perfect for use in wooden structures or with steel rods, provided that the end of the rods that converge at the node are equipped with a plate that has been drilled appropriately.

The intention is to produce a single typology that can join any spatial lattice of rods, liberating the designer from any formal a priori restrictions. The connection joints that are currently available on the market often only permit orthogonal connections or those with predetermined angles—for example, 45°—thus establishing strict formal limitations to the configuration of architectural structures.

For explanatory purposes, it has been decided to consider spatial lattice of rods with both orthogonal and 45° intersections and those with free angles. This lattice allows us to analyse nodes with valence, that is, the number of converging rods, variable between 3 and 8, and to test the efficacy of the method, even in cases where the converging rods come from the same half-space. Other studies show that, in this situation, the connection between the rods is more critical for smoothing algorithms [3].

Below, the essential phases in the algorithmic modelling of a connection joint will be described, with the aim of explaining the logical sequence of the process (Fig. 3.5).

The structuring of a design process in algorithmic terms—that is, according to a logical sequence of instructions that allows the final form to be generated based on geometric-mathematical inputs—had the advantage of obtaining flexible solutions to the different design conditions and can therefore be optimised by means of genetic algorithms. The algorithm has allowed for the creation of a cohesive relationship system between the parts of the design, in which the changes are propagated on every level.

Once the rod network has been defined during the design phase, the points of intersection for the lattice of the rods must be identified and, for each of these, we must identify the other vertices of the lattice connected to these via a rod. This preliminary phase allows us to attribute the respective rods to each node and the relative directions of vectors leaving the node. This is essential for organising, in a

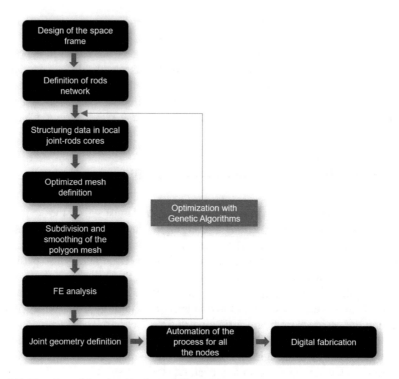

Fig. 3.5 Flowchart of the algorithmic modelling of the connection joint

hierarchical fashion, the data into local working nuclei, which are constituted by the node and the rods connected to it (Fig. 3.6).

Working on a single node, we will proceed by defining a polygonal mesh that is composed of a central nucleus and a varying number of branches that have the same direction as the rods converging at the node.

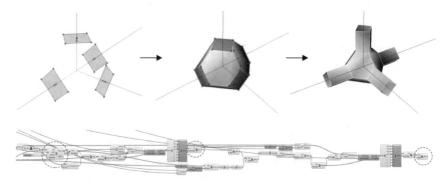

Fig. 3.6 Definition of the simplified polygonal mesh for a node with the rods connected to it

The next phase consists of a process of subdivision and smoothing of the mesh. The order in which the two operations are carried out determine the final form of the connection joint. In a mesh with a low level of subdivision (a low number of faces), the effect of a smoothing algorithm will be more visible and will lead to a significant variation in the volume of the object; in the case of a mesh with a high level of initial subdivision (a high number of faces), the smoothing will have a marginal effect and will only be noticed in the areas around the edges (Fig. 3.7).

By working on the number of iterations and the order in which these two operations are carried out, it will be possible to obtain connection joints with significantly different forms and mechanical performances.

The choice of the best solution has been carried out by means of an optimisation process that uses the genetic algorithm known as *Galapagos*. We needed to identify a *fitness function* that was representative of the static behaviour of the connection joint when under pressure from the rods of the structure to which it is connected. We have chosen to use elastic energy (deformation energy) as a parameter, which is a function of the acting forces and the shifts experienced: the goal of the optimisation process is to identify the configuration of the connection joint, which, under the action of a given set of forces acting on it, has the least elastic energy, that is, it undergoes fewer deformations (Fig. 3.8).

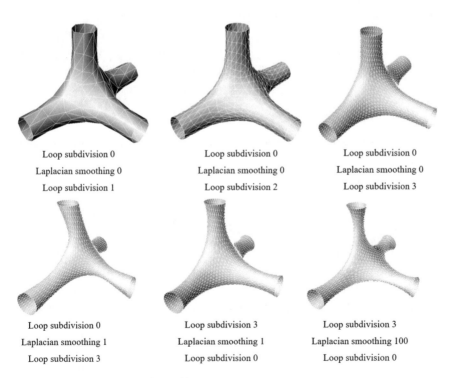

Fig. 3.7 Mesh-subdivision and smoothing process

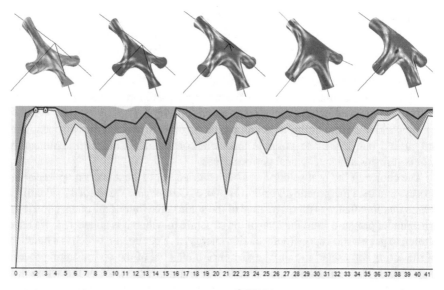

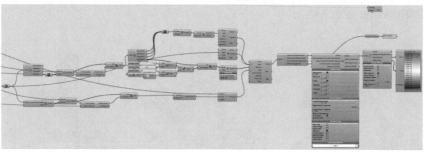

Fig. 3.8 Progress of the optimization process through the use of the *Galapagos* genetic solver in conjunction with the FEM analysis of the connection joint (Elastic energy respectively: 1034 J, 239 J, 213 J, 125 J, 58 J)

The structural analysis, therefore, has been established by attributing fictitious loads to the structure, which stem solely from the self-weight of the rods. The rotation at the internal nodes is restrained and the structure is fixed at the base. Loads, constraints, materials and sections of the rods can be updated at any time; as a result, it was decided to focus on a case that was as theoretical as possible, instead concentrating on the process (Fig. 3.9).

The first part of the algorithm developed makes it possible to identify the forces that act on a specific node, based on analyzing the entire structure. The software used for this analysis, *Karamba*, allows us to identify the forces at the end of each rod; therefore, it will only be necessary to identify the rods that converge at the node and, for each of these, determine whether the end connected to the node is the initial end (i) or the final end (f).

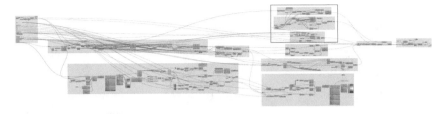

Fig. 3.9 Visual algorithm for the subdivision, smoothing and boolean operations phases

Once the force vectors at the end of the rods connected to the node have been calculated, we must invert and apply them to the ends of each branch of the connection joint and verify the balance of the forces acting on the node, the vectorial sum of which must be zero.

We then proceed with the FEM analysis of the polygonal mesh surface, which allows us to identify stress distribution, displacements and, as a result, elastic deformation energy. The genetic algorithm will be established to find a solution that minimises the selected *fitness function*, that is, elastic energy.

The parameters that determine the varying form of the connection joint are controlled—as seen above—by the relationships between the subdivision of the mesh surfaces and smoothing.

The algorithm will make it possible to test dozens of possible combinations, while maintaining an identical final level of subdivision for the polygonal mesh surface.

A series of Boolean subtractions will make it possible to define the housing for the plate and the bolts at this point (Figs. 3.10 and 3.11).

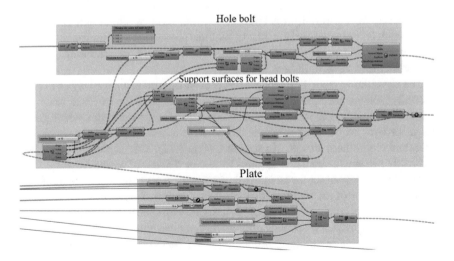

Fig. 3.10 Visual algorithm for the subdivision, smoothing and *Boolean* operations phases

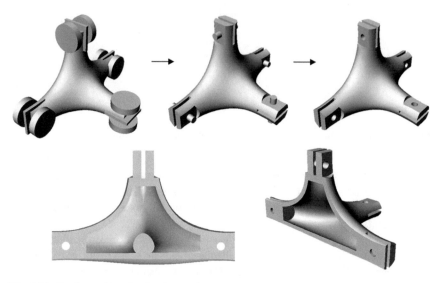

Fig. 3.11 Boolean subtraction for the creation of the connection joint's final geometry (top). View of the hollow section of the connection joint (below)

A version of the hollow internal connection joint was also produced to reduce the amount of material needed and, as a result, the weight of the piece. This solution was adopted for the rapid PLA prototyping with FDM technology, as described in Chap. 3.5. By cutting the connection joint with a vertical section plane, it is possible to visualise the inside of the piece. The internal cavity was obtained again by means of a *Boolean* subtraction from the entire connection joint.

However, the industrial production of a metal, steel or aluminium connection joint, which is completely hollow inside, requires production techniques, such as the laser sintering of metallic powders (DMLS—Direct Metal Laser Sintering), that are more costly than CNC milling of the piece.

The algorithmic modelling process has been structured in such a way as to automatically generate a connection joint at each intersection of the lattice of roads that have been supplied as an input. By connecting a timer to the algorithm, which generates an ordered sequence of numbers, it is possible to generate all of the connection joints and report them on a grid in sequence (Fig. 3.12). This step makes it possible to directly prepare the connection joints for the next phase of production.

The timer has been set for an interval of 1 s, which is a sufficient amount of time for the calculation of each connection joint. The potential of parametric modelling, as can be seen in this application, is precisely the ability to regenerate all of the necessary components in just a few seconds: infinite design possibilities can be tested before production, without investing additional time and money.

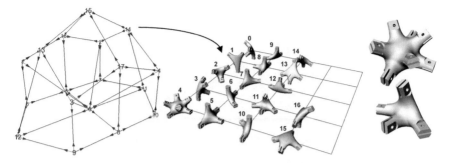

Fig. 3.12 Automatic generation of the 17 connection joints for the test structure and their repositioning on a production plan

3.5 Digital Fabrication of a Reticular Spatial Structure Prototype

In order to validate and test the defined technological solution, it was decided to create a standard module of a full-scale spatial reticular structure with 8 culms of the Viridis species used as rods (for the features of this material, see Chap. 2, Sect. 2.1), 5 different types of connection joints and 16 connective plates (Figs. 3.13, 3.14 and 3.15).

In paragraph 3.4, we described the three-dimensional modelling and control phase of the geometric shape of the connection joint component and the connective plate. The connection joints and plates were printed with FFF additive technology.

From the digital model to the final production of the components, a procedure was followed; this can be broken down into a logical sequence that spans from the verification of the model to the actual printing phase.

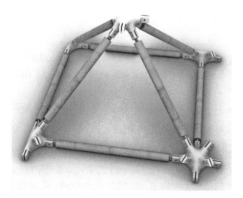

Fig. 3.13 Prototype structure module

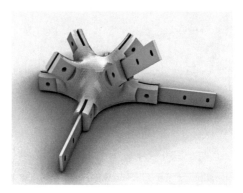

Fig. 3.14 Connection joint-plate model

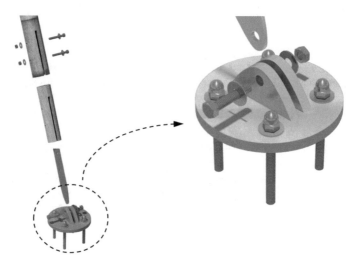

Fig. 3.15 Detail of the ground anchor

The CAD/CAM process requires that the digital model be carefully controlled and managed in order for implementation to occur. This must satisfy a series of geometric/topological requirements [4, 5].

The workflow for producing the prototype is outlined below.

1. **Checking the polygonal mesh model**

The NURBS mathematical solid modelling of the design components must generate multi-surface models that are watertight and have no self-intersections, overlapping edges or overlapping faces. These geometric-topological errors are often linked to Boolean subtraction operations and unions between un-connection jointed elements. To proceed to the subsequent steps and use management software for the 3D-printing process, it is necessary to tessellate the geometry that converts the mathematical model into polygonal mesh (Fig. 3.16) [6].

The mesh models that are obtained as output from the parametric algorithm (described in paragraphs 3.3–3.4) must be sufficiently verified before moving to subsequent phases of the rapid prototyping process.

The polygonal network that is generated could present problems or topological errors linked to the tessellation of the mathematical surface, which could make the components impossible to print.

Understanding the formal management methodology for the parametric design and the digitalisation of the model are essential for the entire production cycle, representing the fulcrum on which the entire process rests. In general, correct solid modelling makes it possible to avoid many of the problems outlined below (Figs. 3.17 and 3.18).

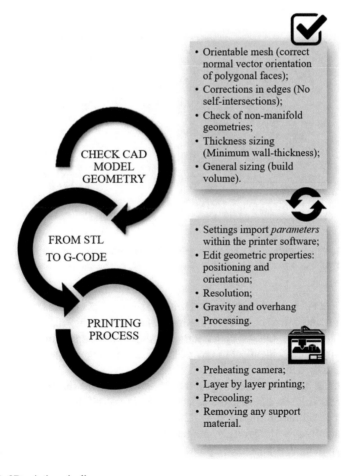

Fig. 3.16 3D-printing pipeline

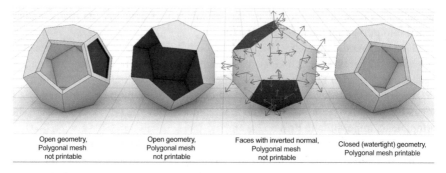

| Open geometry, Polygonal mesh not printable | Open geometry, Polygonal mesh not printable | Faces with inverted normal, Polygonal mesh not printable | Closed (watertight) geometry, Polygonal mesh printable |

Fig. 3.17 Topological errors that make it impossible to print the polygonal mesh

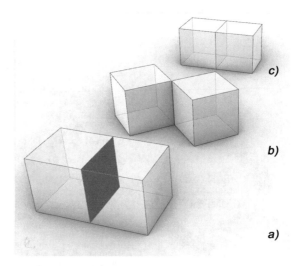

Fig. 3.18 Topological errors that make the polygonal mesh non-manifold and thus impossible to print: **a** overlapping; **b** edge shared by more than two faces; **c** printable, manifold geometries

- They must not have open edges, ensuring that the object is closed and has its own internal volume. In fact, an open mesh geometry has no thickness and, as a result, cannot be printed. Only a collection of juxtaposed meshes—in which the edge of each polygon is identical to that of the adjacent polygon—allows for a closed volume to be defined.
- There must be no manifold edges, that is, edges that are shared with more than two faces. This would cause the object to intersect itself. In order to avoid errors, the numerical modelling must be "2-manifold", that is, if more than two faces share an edge, an ambiguity is created. In fact, if more than two faces share an edge, anything that surrounds a point and relies on said edge could belong to four faces.
- There must be no discontinuities or gaps in the polygon network.

- Each polygonal face must have the perpendicular part of its surface turned towards the outside of the solid volume. The correct course of the perpendicular part determines the consistency between the internal and external parts of the object's surface envelope, which makes it easy to tell if the model is concave or convex. Inverted perpendicular vectors of two adjacent faces lead to printing errors.
- Depending on the material and the type of printing technology that is used, dimensional limits, such as the minimum thickness or maximum size of the object, must be adhered to. The walls of the surface envelope must have a minimum thickness, the so-called "wall thickness". It is generally recommended not to go beneath one millimetre of thickness.

2. **Conversion from the CAD model to the G-Code**

Once the verification has been carried out and any problems have been resolved, the phases that are inherent to the prototyping of the object finally began.

The model was thus exported from *Rhinoceros* in *stl* format (Standard Triangulation Language), the famous format that makes it possible to communicate between CAD modelling softwares and printing-management softwares (other commonly used formats include.*obj;.ply; 0.3mf;.amf;.pov*).

The exportation in stl format makes the surface of the object in triangles discrete and stores the data in the form of spatial coordinates *[x, y, z]*, which represents the vertices of the triangles and vectors that represent the perpendicular parts of the faces.

The file in *stl* format was thus imported into the *Simply3D* professional software, which makes it possible to establish the printing parameters and define the critical areas where the support material will be stored, as well as optimising the printing process in terms of quality, final rendering and time.[2] There are many softwares available; some are free (the most famous include *Cura*, developed by UltiMaker and *Repetier-Host*, developed by Hot-World GmbH), others open source (*Slic3r*) or paid. These are commonly defined as *slicers* and make it possible to optimally configure the numerical variables in relation to the print configuration that is intended for use and the type of printing that will be used.

The protruding parts, or those with accentuated inclinations (usually 45° is the safety limit), require a supporting "scaffolding" during the printing phases, in order to avoid structural collapse. Through the slicing operation, the plastic filament is melted down and deposited on the platform in overlapping layers.

The material used for the 3D printing of the components is polylactic acid (PLA), a biodegradable, thermoplastic polymer of vegetable origin (tapioca, corn, sugar cane or potato) that is non-toxic and insoluble.

The machine used for the 3D printing of the components is a R3D 335/40 produced by the Palermo-based company R3Dit, an open-source Arduino RepRap design.

The Cartesian 3D printer has a heated glass and aluminium printing plate with a print volume of 250 × 350 mm, which increases in height to 400 mm, making it unique as a 3D printer in the entry-level sector. Its structure is made of aluminium,

---

[2] Further information available at https://www.simplify3d.com/.

it has a standard 0.4 mm nozzle and its vibrations are minimised thanks to the movements of the carriage being limited to the *y*-axis (Fig. 3.19).

We can distinguish 3 phases in the processing of the *stl* file, in order to prepare it for printing, so that the virtual volumetric structure is translated into a unique language (G-Code format) in a series of physical instructions that are required by the machine for performing the material construction of the model ("Slicing" operation). The 3D printing workflow and the establishment for the most important parameters are described below.

Positioning and orientation

The first operation performed in the program is the spatial positional and orientation (rotation and movement along the coordinate axes) of the object along the printing plane. This is managed manually and relies on the experience of the operator; although there is typically no ambiguity about the orientation of the piece, in the event of connection joints with branches in various directions, which do not have a preferred support plane, we must consider the static equilibrium of the pieces during the entire printing process and the amount of supporting material that is needed in various potential configurations, which, at the end of the process, will be eliminated (Fig. 3.20).

The maximum angle of protrusion (maximum overhang angle) of the structural parts of the component to be printed has already been discussed; in relation to the

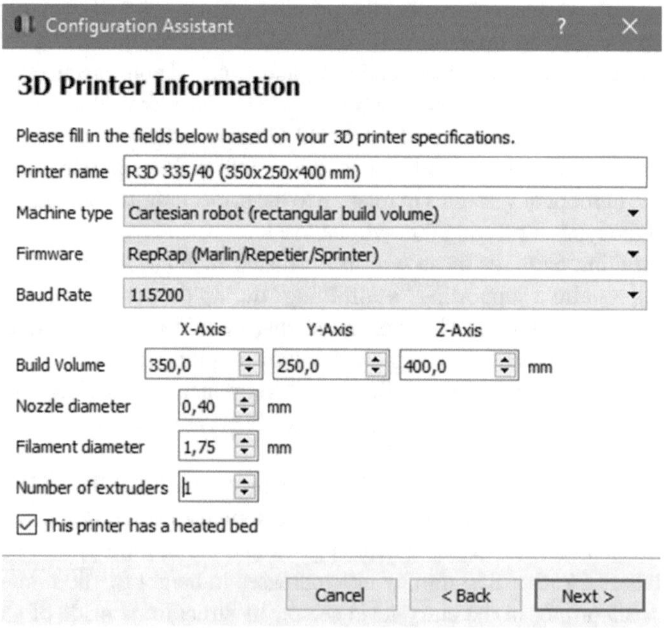

Fig. 3.19 Settings print parameters

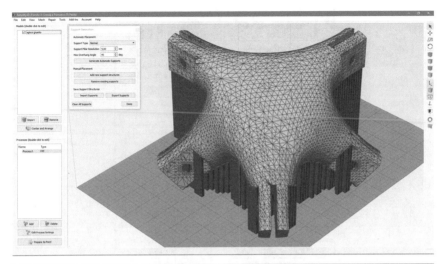

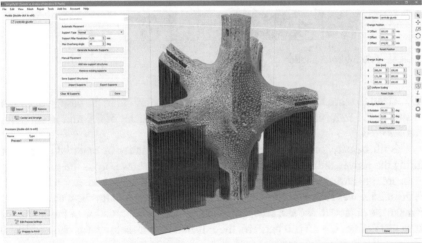

Fig. 3.20 Layout in the printing plate (software used simplify 3D 4.0)

structure of the object within the work area, the printing of the support material could negatively impact production times.

Settings

The three most important parameters during the subsequent phase for establishing printing settings are "layer height", "fill density" and "Outline/perimeter shells". These significantly influence printing times, costs and the quality of the final object (Fig. 3.21).

The layer height is the thickness of the material layers, in direction *z*, which will be progressively deposited and are compatible with the diameter of the nozzle. Higher

Machine Setting	Basic/Advanced parameters
Machine type	Cartesian robot
Nominal resolution	X e Y: 0,015 mm - Z: 0,39 micron
Layer height	0,1 mm
Shell number/outline	3
Top/bottom solid layers	4
Fill Density	25 %
Printing temperature	210 °C
Build Plate temperature	60 °C
Filament PLA Diameter	1,75 mm
Nozzle Extruder Diameter	0,40 mm
Flow	100%
Enable retraction	active
Print speed	90 mm/sec
Travel speed	120 mm/sec
Average print time	65 h

Fig. 3.21 Printing configuration

values lead to more rapid printing, but with lower resolution. In the case in question, a value of 0.1 mm (high resolution) with a nozzle diameter of 0.4 mm in extrusion was selected.

The fill density, on the other hand, acts on the quantity of materials deposited during the movement on each *xy* plane and determines how porous the material is.

In order to ensure structural rigidity of the connection joint geometry, a fill density value of 25% was selected, while a value of 100% was chosen for the plates, in order to make the objects dense and compact. In terms of the configuration of the geometric form that defines the fill (fill pattern), the full-honeycomb setting was chosen, as this is the most effective in structural terms.

The third setting manages the characteristics of the shell, that is, the thickness of the shell, by setting the number of perimeter print lines that comprise the walls in a horizontal direction. More specifically, the printed models have three perimeter lines—one external and two internal—in order to ensure the shell is rigid.

Processing the G-Code

At this point, the software processing begins. This breaks down the model into a series of overlapping layers ("slicing" operation) and defines the parts where supporting materials will need to be applied. Slicing is a geometric-mathematical operation of layer discretisation, which involves breaking the digital model down into a series of parallel layers or slices. In IT terms, this translates into the production of a file in G-Code format; this consists of an ordered sequence of instructions (managing the print speed of various layers, the temperature of the plate and nozzle, the motors, the release flow of materials) and spatial coordinates that define the process that the extruder of the machine will be forced to follow during the 3D-printing phase (Fig. 3.22).

To ensure that the process is successful and that all of the layers have been processed correctly, it is possible to visualise the progression of the levels to various degrees.

The software preview displays the layers by differentiating the process of the extruder for generation by colour: the external perimeter walls; the upper and lower walls; the perimeter thickness lines; the support structures; the movements; any *brim* or *raft* to ensure the plate adheres more effectively. In particular, the *brim* represents an extension of the first printing layer, which comprises a number of extra, user-defined lines that improve the application of the initial extrusion layer, improve adhesion, avoid deformation of the physical model and simplify the operation of removing the piece at the end of the process.

3. Printing the object

Once the G-Code processing phase has been completed, it is possible to transmit the information to the printing machine. Before the extruder nozzle starts to deposit the material, the plate and the nozzle are pre-emptively heated, for a period of approximately 2 min, in order to avoid the PLA shrinking or deforming. It is extruded at a temperature of approximately 200 °C. The printing process for the 5 types of connection joints that were determined required an average printing time of approximately 60 h for high resolution (Fig. 3.23).

At the end of the process, the plate was progressively cooled down to room temperature and the motors and fans were automatically stopped once the extruder was placed in its final position. The final operation involves the mechanical removal of the supporting material and the potential removal of roughness or imperfections using abrasive paper (Figs. 3.24 and 3.25).

At this point, it is possible to assemble the connection joint with the end of the rods. This, in fact, is the only phase that is carried out on-site, since the preparation of the bamboo rods and the connection joints takes place in a laboratory.

The operation consists, for the most part, of tightening the bolts and does not require specialised workers (Figs. 3.26, 3.27, 3.28 and 3.29).

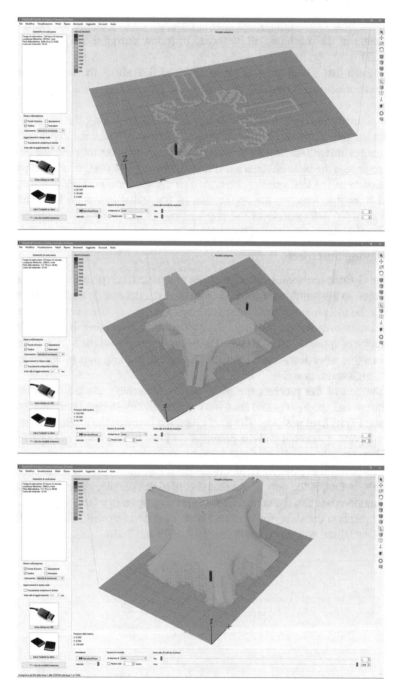

Fig. 3.22 Result of the phases for slicing, calculating support materials and processing the G-Code. Visualisation of the layers at varying levels (software used simplify 3D 4.0)

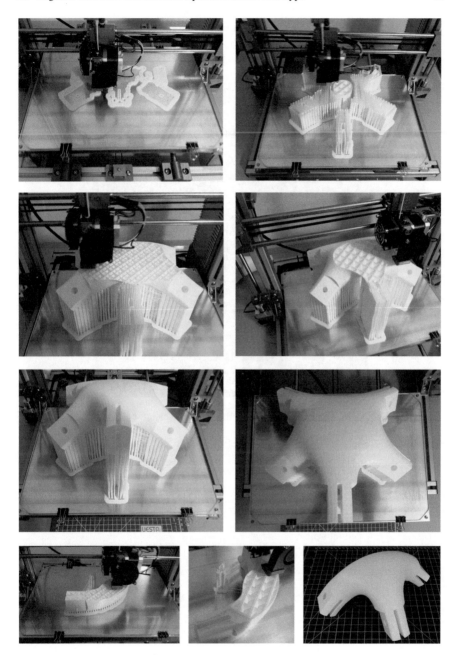

Fig. 3.23 Typical connection joint printing

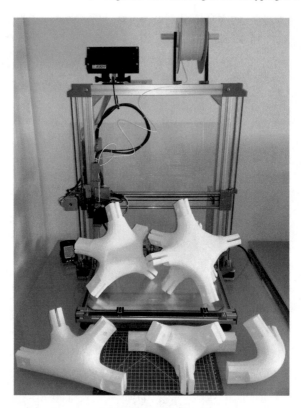

Fig. 3.24 Five 3D-printed connection joints for the structure module

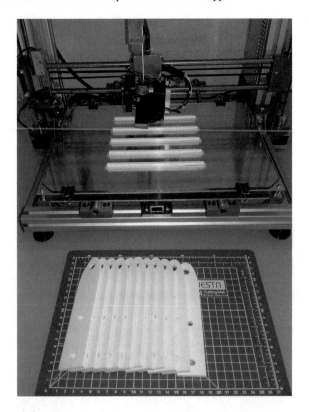

Fig. 3.25   Plate printing

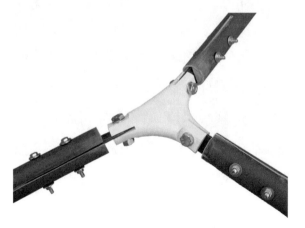

Fig. 3.26   Assembly of a typical connection joint with three bamboo rods

Fig. 3.27   Assembly of connection joints and bamboo rods of a structure module

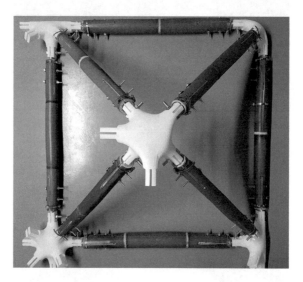

Fig. 3.28   Top view of connection joints and bamboo rods of a structure module

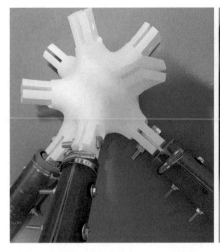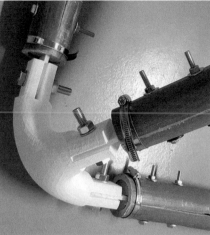

Fig. 3.29 Details of the connection joints of a structure module

References

1. Arce-Villalobos OA (1993) Fundamentals of the design of bamboo structure. Phd dissertation, Eindhoven University of Technology
2. Janssen J (1981) Bamboo in building structures. Phd dissertation, Eindhoven University of Technology
3. Schneider R, Kobbelt L (2001) Geometric fairing of irregular meshes for free-form surface design. Comput Aid Geom Des 8(4):359–379. https://doi.org/10.1016/S0167-8396(01)00036-X
4. Gibson I, Rosen D, Stucker B (2015) 3D printing, rapid prototyping, and direct digital manufacturing. Springer
5. Popescu G (2007) Digital material for digital fabrication. Master Thesis, Massachusetts Institute of Technology
6. Tedeschi A (2014) AAD algorithms-aided design. Parametric strategies using Grasshopper. Le Penseur Publisher, Potenza

Chapter 4
Structural Patterns for Free-Form Surfaces

Abstract In the previous chapter, we defined the design process for connective joints in spatial reticular structures with rods made from bamboo culms. The following chapter, on the other hand, analyses a series of themes related to the design and optimisation of a free-form structure. Once the form of an NURBS mathematical surface has been determined, we investigate the main techniques for subdivision of discrete elements with dimensions adapted for manufacturing and assembly (paragraph 4.1 *Digital representation techniques, NURBS and Polygon Mesh*; paragraph 4.2 *Optimisation of surface tessellation*). The discretisation method of a planar surface based on Voronoi diagrams is detailed further (paragraph 4.3 *Natural patterns, the Voronoi diagram Particle-Spring System and definition of the structural grid*) by optimising the distribution of the cells in the pattern with form-finding techniques, in order to satisfy structural requirements (the algorithm used is the *Particle-Spring System*, by means of the *Kangaroo* plug-in for Grasshopper, which was developed by Daniel Piker). In the last Sect. 4.4, *Definition of the numerical model and structural optimisation through the Galapagos evolutionary solver*, the diagram is applied to any free-form surface by iterating an algorithmic process to develop the best geometric-spatial configuration that satisfies certain threshold values.

4.1 Digital Representation Techniques: NURBS and Polygon Mesh

In architecture, but also many other fields, design often uses nature as a source of inspiration. If we focus on the infinite points of departure that it offers us from a structural perspective—from a microscopic to an extensive scale—we notice that nature has no typical geometries that always satisfy the same rules. This stems from the fact that natural structures have adapted to irregular and non-uniform forces, optimising and balancing a general framework that best responds to the context in which it occurs [1].

On the other hand, throughout the history of architecture, the search for a serialisation of elements and the consistent use of straight lines and orthogonal connections has always been a priority. This was done to make construction simple and because there were no digital tools to control the geometric-formal dynamic in relation to varying environmental conditions.

Starting a few decades ago, the continuous evolution of computing tools and methods made it possible to enhance design and engineering production processes.

Virtual representation, *free-form* surface-modelling techniques and numerical-control manufacturing, with their intrinsic dynamic and interactive capabilities, have significantly expanded and enriched the repertory of geometric forms, creating innovative design skills and creative languages. For some examples, see: *deconstructivism*; the more recent *Blob Architecture*; "file to factory", a term coined by the *Objectile* group or the "Non-standard architectures organised by the Centre Pompidou" exhibition held in Paris many years ago, in 2004.

Deconstructivism is an architectural movement that stems from an exhibition hosted in New York in 1988, which showcased designs by Frank O. Gehry, Daniel Libeskind, Rem Koolhaas, Peter Eisenman, Zaha Hadid. The rules on which this architectural style is founded aim to deconstruct what has been built. The message it communicates lies in an architecture that is free from geometry (geometry, as it is understood based on Euclid's teachings), a type of non-architecture that is expressed in the plasticity of its volumes of shapes, which are complex, unstable, disjointed, fragmented and deformed, as well as cut, asymmetrical and free from the traditional aesthetic principles. In summary, it offers a new view of the built environment and architectural space, which, by deforming geometry, describes innovative sculptural forms.

Blob architecture, on the other hand, started in 1995. The term was coined by an American architect, Greg Lynn, who created the *BLOB* design software (Binary Large Objects). This makes it possible to amend the algorithms of a system by predicting the evolution of forms, based on the pressure of external forces on the surfaces.

Contemporary architectural creations—by Peter Eisenman, Zaha Hadid, Norman Foster, Massimiliano Fuksas, Renzo Piano, Toyo Ito and Frank Owen Gehry, to name just a few—show innovative contributions and experimentations in the use of computers for design, tecnique and the simultaneous management of shapes, construction technologies and the production of all the components in a building.

A fundamental theme, which must be tackled for a conscious and contemporary approach, is that the geometric complexity of *free-form* shells cannot be managed separately from structural considerations: it is from the shape that the mechanical performance of the structure derives (Fig. 4.1) [2].

It is clear that static and technological considerations should be taken into account from the very beginning of the design process of the shape.

However, in many cases, by analysing the process that led to the production of numerous free-form structures, we can see the absence of real integration and optimisation between architectural design and structural design. The supporting structure remains predominantly based on the traditional frame model, while the *free-form* shell is simply a supported element [3].

The concept of optimisation can be also be expanded to issues that do not concern the structural aspect; for example, the objective of the optimisation may concern

Fig. 4.1 Ondulated roof at the Louvre Museum's Department of Islamic Arts, 2005–2012, Paris (Mario Bellini Architects with French architect Rudy Ricciotti)

the search for optimal shading, acoustic performance or other problems that may be expressed in the form of numerical functions that are to be minimised or maximised.

Formal research by means of optimisation techniques is referred to as computational morphogenesis.

As is noted in the literature, it is possible to define two possible approaches to the problem of optimising *free-form* design.

The first presupposes a posteriori optimisation (the *problem of approximation*), in which the form is considered as a piece of initial data and, as a result, the surface is broken down into panels through differential mathematical and geometric techniques.

For example, this is the case with *Admirant Shopping* by Massimiliano Fuksas, which was produced in Eindhoven between 2003 and 2010 [4]. The shape of the shell, an ellipsoid that is locally distorted as an expression of the architect's artistic vision, has already been defined, and the engineering process does not make significant changes to the initial geometry, thanks to sub-dividing algorithms (Fig. 4.2) [5].

The second approach, on the other hand, pertains to a priori optimisation (*design problem*), in which specific requirements have already been considered during the design of the shape.

In this case, the form will be the result of an optimisation process, such as, for example, *form-finding* (examples are found in Chap. 1, *Form-finding and digital simulation techniques*). This method places geometric or static considerations at the core of the creative process: the form is defined based on the forces that act on it.

Both approaches will be used as the study continues: the first, in order to validate the efficacy of a discretisation method based on Voronoi diagrams; the second, for the formal definition of an exhibition pavilion.

Graphics computer softwares chiefly use two representative techniques to describe any three-dimensional geometry: mathematical representation, which uses NURBS

Fig. 4.2 Admirant shopping, Eindhoven, photos of the building and details (reproduced from Lee 4)

curves and surfaces; numerical boundary representation (*boundary representation, b-rep*), that is, approximation through polygonal surfaces, also known as *mesh*.

Non-Uniform Rational Basis Splines (NURBS) are mathematical representations of a geometry, which make it possible to accurately define any form.

A NURBS geometry is predominantly described by a mathematical formula that relates its various components: the degree, control points and nodes.

The geometry can be amended by acting on the control points or varying the relevant weights: an increase (or decrease) in the weight attributed to a control point will lead to the curve approaching (or distancing itself from) the curve of the control polygon (defined from the points $\{P_1, P_2, ..., P_5\}$) (Fig. 4.3) [6].

Generally, in the case of NURBS geometries, each point on the curve or surface is not identified using Cartesian coordinates in the global reference system, but, instead, with a local parametric reference system.

For example, in the case of a curve, each point will be identified solely with a parameter, t, the domain of which will be "reparametrized" in the range between 0 and 1 (Fig. 4.4).

In the case of a surface, each point will be identified with a pair of values, u and v, in a two-dimensional domain ranging from 0 to 1 in both directions (Fig. 4.5).

Conversely, if we consider the topological structure of a *mesh* surface, this is not a single geometric entity, but instead comprised of a collection of adjacent polygons that approximate the overall form. Each *mesh* is defined by an established number of vertices, edges and a face and is typical triangular, quadrangular or an *n-gon*.

From this, it is possible to deduce that the amount of information required for the NURBS representation of a geometric element is significantly less than the amount of information needed to represent the same geometry through *mesh* approximations.

The degree of approximation involved in the polyhedral model will be proportional to the number of polygons that comprise the geometry.

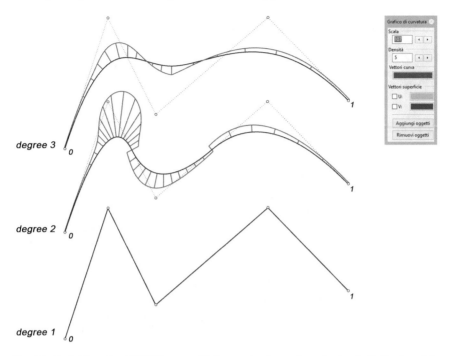

Fig. 4.3 Definition of an NURBS curve with 5 control points and varying levels. Indication of the curvature graph

Despite the need to manage a greater amount of data, the computational resources needed to digitally process a *mesh* are significantly fewer than those needed for NURBS.

In the case of architectural surfaces in the compositional-design context, or a digital survey, or in CAD/CAM applications, the *mesh* models are an effective representation tool, since they make it possible to establish a parallel between: edges, which will comprise the rods of the structure; vertices, which will comprise the joints; faces, which will, in turn, eventually comprise the panels.

These vertices are identified, in their spatial position, by means of Cartesian coordinates; the borders/edges define the connection logic between the vertices, which directly influence the orientation of each face (direction of the normal vectors to the face) and the regularity of the overall *mesh* surface (that is, solid models that are free from ambiguity) [6].

Many of the topological errors related to numerical polygonal-modelling processing typically stem from the application that tracked the tessellation, or they can be attributed to the operator. As was better outlined in Chap. 3.5, *Digital fabrication of a reticular spatial structure prototype*, the most common problems relate to: the discontinuity of the polygon network; the non-homogeneous direction of the normal vectors; the ambiguity of belonging in terms of edges with many faces (non-manifold representation); the presence of overlapping co-planar faces.

A weighted NURBS curve

How do numeric weights correspond to physical weights?

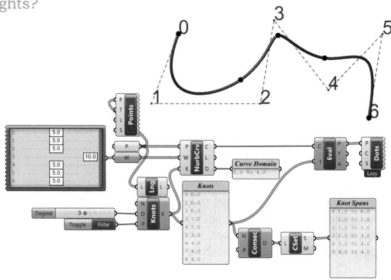

Fig. 4.4 Relation between control points and weights of a NURBS curve

4.2   Optimisation of Surface Tessellation

The chief problem that must be tackled when working with complex structures—as is the case with *free-form* structures—is the subdivision of the surface into discrete elements (discretisation), which define the form in the most accurate manner possible.

This is a highly interesting multi-disciplinary subject, one that is subject to continuous experimental research that brings together mathematics, geometry and technology; this is especially relevant in recent years, thanks to the development of new production processes.

In order to create a structural grill on any surface, the latter is broken down into discrete cells made of polygonal *mesh*; these are typically triangular, quadrangular or hexagonal in shape.

The subdivision into triangular *mesh* allows a strong degree of approximation of a surface, its main advantage being the ability to solely obtain flat faces, given that only one plane passes for three non-aligned points.

Furthermore, the non-deformable triangular coatings often ensure a more stable structural response. However, they present some critical issues, such as low transparency in the case of glass shells and an increased structural weight.

In the case of quadrilateral *mesh*, the complexity of the nodes decreases, since each of them has only 4 rods and not 6, as is the case with triangular mesh.

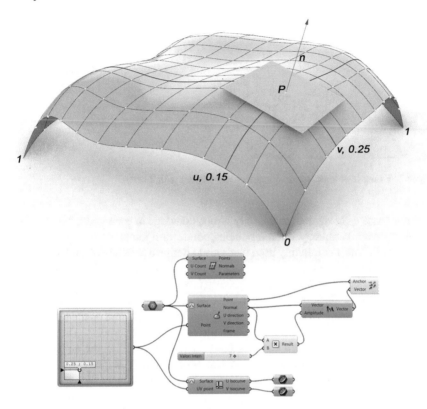

Fig. 4.5 Definition of a NURBS surface with two uv iso-parametric curves highlighted for point P ($u = 0.15; v = 0.25$). Representation of the tangential plane of the surface at point P and the normal vector

Quadrilateral *mesh* does not ensure that faces are planar (quadric surfaces and hyperbolic paraboloid portions are obtained), except for certain special types of surfaces, such as translational or rotational surfaces, in which *Planar-Quad meshes* (*PQ Meshes*) are obtained via geometric methods (Fig. 4.6) [7].

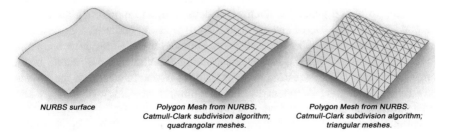

NURBS surface

Polygon Mesh from NURBS.
Catmull-Clark subdivision algorithm;
quadrangolar meshes.

Polygon Mesh from NURBS.
Catmull-Clark subdivision algorithm;
triangular meshes.

Fig. 4.6 Diagram of a surface comprising quadrangular or triangular mesh (Grasshopper, Weaverbird plug-in)

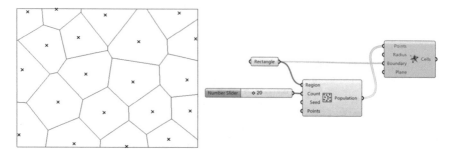

Fig. 4.7 Example of a two-dimensional voronoi diagram with 20 cells

In the case of hexagonal *mesh*, only 3 rods converge in each node; this condition leads to greater simplicity when producing connection joints.

One of the most commonly used methods to define a subdivision pattern involves using a "uniform" grid based on the local *u-v* parametric coordinate system. This allows a good approximation of the source surface, without having to consider aspects relating to the serialisation of the elements, since both the length of the rods and the dimensions and shape of the panels will progressively vary based on the surface trend (Fig. 4.7).

Other techniques—which will not be discussed in detail in the context of this study—involve the use of algorithms and programming languages, in order to produce panels or rods that are more regular in terms of size and, as a result, reduce production costs.

For example, iterative algorithms, which are based on the theory of *Chebyshev nets*, make it possible to obtain a network of equidistant points along the surface and, thus, rods of the same length [8].

4.3 Natural Patterns, the *Voronoi* Diagram (Particle-Spring System) and Definition of the Structural Grid

Voronoi diagrams are named after the mathematician Georges Voronoi (1868–1908), who came up with a general mathematical definition, and represent a subdivision method for two-dimensional spaces, according to proximity criteria.

These diagrams are sometimes defined as Dirichlet tessellations, after a German mathematician who studied some applications of them during the middle of the nineteenth century. They are used in many fields of research, which, on the surface, have little in common: astronomy, urban planning, meteorology, physiology and, more recently, computational geometry [9].

Voronoi diagrams, in two-dimensional cases, make it possible to break a plane down into a number (n) of cells, equal to the number of points $P_i = \{P_1, P_2, ..., P_n\}$ assigned on the plane.

The lines that define the diagram have the property of being equidistant from the two points closest to them. This implies that the cells are convex [10].

In other words, for every point, Q_j, belonging to a cell with P_i as its centroid, the distance P_iQ_j is less than the distance of Q_j from any other centroid assigned on the plane.

What is perhaps even more interesting is that *Voronoi* diagrams can be found at every level of nature. In the process that led to the formation of this kind of *pattern*, a spontaneous process of optimisation can be noted: in nature, everybody that is subject to an external or internal field of force tends to reach a state of equilibrium, minimising the potential energy it possesses (Figs. 4.8 and 4.9) [11–14].

By analysing the *Voronoi* diagram, which has been mathematically crafted using the proximity criterion as the sole limiting condition, it is clear that this diagram does not have the appropriate features for use as a structural *pattern*. The rods are of very different lengths, while some have infinitesimal dimensions; furthermore, the

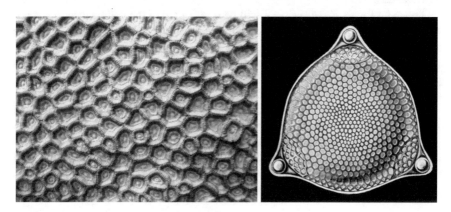

Fig. 4.8 Left: coral. Right: diatom (micro-alga)

Fig. 4.9 Left: dragonfly wings. Right: cracks in an arid landscape

irregular distribution of cells makes this kind of subdivision less appealing from an aesthetic point of view.

Therefore, it is necessary to optimise the *Voronoi* diagram, "relaxing" the mesh in order to create a solution in which the potential energy of the system is minimised. This is the very thing that occurs in nature, in terms of the balance between internal and external forces. To this end, the *Particle-Spring System* was used [12].

Each node of the mesh is considered to be a hinge, while each rod connected to it can be deformed axially in accordance with *Hooke's Law*. By assigning a "resting length" equal to zero to the rods and extending them—in the same manner as elastic springs—to their respective length L_i in the initial condition (obtained from the *Voronoi* diagram), a system is obtained in which each rod is subjected to a pre-tensioning process.

The extent of the force generated in this way on each rod will be a function of the elongation experienced (ΔL) and the axial rigidity of the material, in accordance with *Hooke's Law*.

The axial forces F_a, F_b, F_c, F_d will be transmitted to the node, where a state of equilibrium must be reached, while taking any external loads applied to the node into account (Fig. 4.10) [15].

Once the equilibrium equations for each node have been obtained, it will be possible to write—in the form of a matrix—a system of equations—from which the displacements can be derived and, therefore, the new coordinates for the nodes.

As a result, the procedure will be repeated by the software until the result obtained as a vectorial sum of the actions transmitted to the node is zero or is in equilibrium with any external load applied to it. During the various iterations, the system will progressively tend towards a state of equilibrium, while maintaining the initial topology of the *Voronoi* diagram.

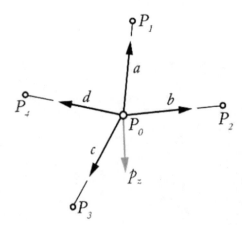

Fig. 4.10 Equilibrium between internal and external forces at P_0 hinge node. The forces F_a, F_b, F_c, F_d are applied to it in an axial direction (reproduced from Adriaenssens et al.)

A configuration will be reached where the distribution of forces is uniform, the potential elastic energy is minimised and, in which, as an indirect consequence, the rod lengths and the polygonal angles will be more uniform.

The development of the algorithm for macro-phases will be described below, with a rectangular surface taken as an example, as outlined in the following flow chart (Fig. 4.11).

The first phase consists of generating a random set of n points, which represent the centroids of the *Voronoi* tessellation. These will have randomly generated u and v coordinates within the surface domain. These points can be used as centroids for geometric tessellation, based on a proximity criterion (Fig. 4.12).

Based on the diagram that is obtained, the geometric entities are transformed into physical entities (Fig. 4.13):

- the lines become deformable rods with resting lengths of zero (resting length = 0) and are subject to a pre-tensioning force that is proportionate to their length; in fact, by assigning a unitary elastic coefficient to all of the rods, the axial force will be directly proportional to the length of said rod, in accordance with *Hooke's Law*. The elastic coefficient of the "spring" has no influence on the final configuration

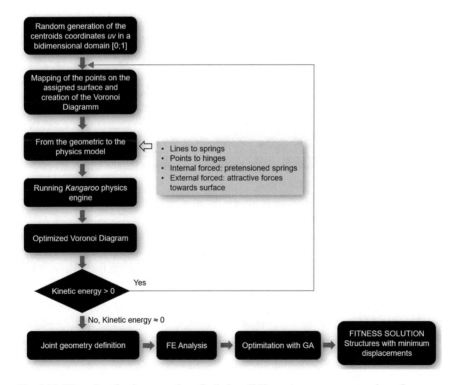

Fig. 4.11 Flow chart for the generation of a "relaxed" *Voronoi pattern* on a generic surface

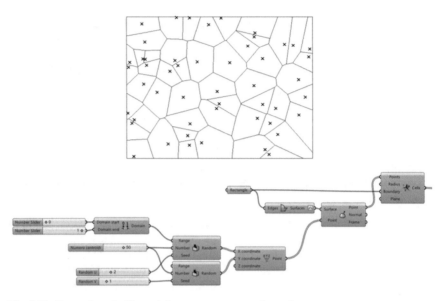

Fig. 4.12 Generation of a *Voronoi* diagram on a rectangular surface

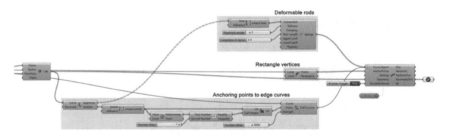

Fig. 4.13 From geometric diagram to physical system (*Grasshopper*, plug-in *Kangaroo*)

of the optimised diagram but only affects the number of iterations needed to reach a balanced configuration.

- the points of intersection become the nodes in which the balance of the forces to which they are subjected are calculated. The curves at the edges and the vertices of the rectangle will be the limitations of the system, ensuring the adaptation of the diagram to the assigned surface.

Once the force system and geometric entity has been assembled and, subsequently, the *Kangaroo*[1] physics engine has been launched, the nodal displacements will be calculated on an iterative basis and, as a result, so will the new configurations of the diagram (Fig. 4.14).

---

[1] *Kangaroo* is a *Grasshopper* physics engine for interactive simulations, optimisation and *form-finding*. It was developed by Daniel Piker. Further information available at http://www.grasshopper3d.com/group/kangaroo.

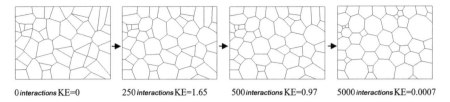

0 *interactions* KE=0   250 *interactions* KE=1.65   500 *interactions* KE=0.97   5000 *interactions* KE=0.0007

Fig. 4.14 Optimisation process of the *Voronoi* diagram

Fig. 4.15 Flow chart for the analysis of the structural pattern

The sequence of displacements describes the motion of particles: they will have a unit mass and an established speed, from which the value of the system's kinetic energy can be derived.

As the state of equilibrium is gradually reached, the extent of the displacements within each iteration decreases and, as a result, so too does the kinetic energy (KE).

When it has reached a value near zero, it will be possible to halt the calculation.

4.4 Definition of the Numerical Model and Structural Optimisation Through the *Galapagos* Evolutionary Solver

The objective at this point is to verify whether the *Voronoi* diagrams can be more or less effective, from the static point of view, for the discretisation of architectural surfaces, according to the following process (Fig. 4.15).

To this end, it was necessary to perform Finite Elements Method (FEM) analyses of the structures in their various configurations.[2]

Since the research aims to investigate the geometric and topological aspects, that is, the various possible configurations of the structural patterns, identical sections and materials have been selected for all of the rods, so this choice does not influence the results obtained.

---

[2] The analyses were carried out within the *Grasshopper* parametric environment by means of the *Karamba* plug-in for structural analysis. Further information available at http://www.karamba3d.com/.

In terms of external constraints, the decision was made to limit the points on the edges of the surface—as is generally the case with *gridshells*—fixing all 6 degree of freedom.

The loads that the structure is subjected to exclusively comprises the self-weight of the rods and a uniform distributed load of 1 KN/m^2.

The only real variable is the subdivision pattern that has been adopted.

Once the FEM analysis has been carried out, we must define the terms for comparing one solution with another. Since genetic algorithms offer the opportunity to find the *optimum* solution by minimising or maximising a given *fitness*, we must identify a function that can represent the overall behaviour of the structure with a single value.

For this reason, the sum of the nodal displacements has been selected as the term for comparison. For each i node, the displacements Δx, Δy e Δz in three directions were calculated and, as a result, so were the resulting vectors:

$$\Delta_i = \sqrt{\Delta_x^2 + \Delta_y^2 + \Delta_z^2}$$

The sum of the n node displacements for the structure will be given by:

$$\sum_{i=1}^{n} \Delta_i$$

It is possible to consider other outputs as a *fitness* function, such as the sum of the *Von Mises* stress calculated at the end of each rod, or the *Von Mises* stress deviation compared to the average value.

The objective of these kinds of optimisation is to ensure an even distribution of stress, avoiding the presence of areas of the structure that are under excessive stress and others that are under very little stress.

Firstly, it is necessary to understand how the "relaxed" *Voronoi* diagram, along with the *Particle-Spring System*, can be applied to any surface and optimised using genetic algorithms as part of a search for a solution that ensures the lowest possible sum of nodal displacements.

In the previous section, a method for obtaining the diagram on a plane was outlined; in the case of a generic surface, the algorithm must be reworked appropriately, even if there are no substantial differences on a conceptual level.

Given that it is impossible to generate the *Voronoi* diagram directly on a non-planar surface, it will first be necessary to randomly generate a series of points on the generic surface and transpose their uv coordinates on the xy plane, in a rectangular surface, within which the *Voronoi* diagram will be generated (Fig. 4.16).

The next step is to remap the diagram obtained on the assigned surface. In this case, this is not simply a remapping of the coordinates from one domain to another. The most important aspect is to respect the connective relationships between the diagram's vertices and the hierarchical organisation of the cell data (points and lines).

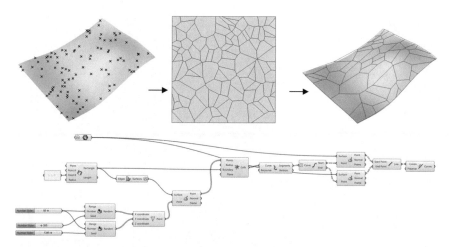

Fig. 4.16 Mapping of surface points on a rectangular plane and generation of the *Voronoi* diagram, which was subsequently remapped on the source surface

The "relaxation" of the diagram will occur on the source surface. In this case, the pre-tensioning forces will trigger the motion of the nodes, which must be sufficiently constrained so that they move along said surface by means of a system of attractive forces. The calculation will be stopped once a balanced configuration has been achieved (Fig. 4.17).

The next phase pertains to the definition of conditions regarding constraints, loads, materials and sections through *FEM* analysis, in order to calculate the nodal diplacements.

In this way, the nodal displacements will be obtained in vector format $v_i = (x_i, y_i, z_i)$; the *fitness* function, which is to be minimised using the *Galapagos* genetic

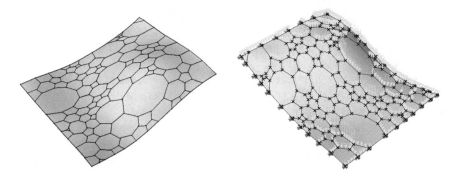

Fig. 4.17 "Relaxing" of the diagram on the assigned surface

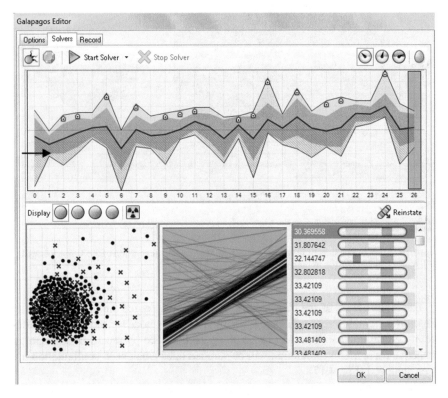

Fig. 4.18 Process for advancing the genetic algorithm

algorithm,[3] will be obtained from the sum of the modules of each displacement vector.

As has previously been mentioned, the uv coordinates of the cell centroids are randomly generated; the use of a genetic solver will make it possible to vary the position of each point within the established domain, thus allowing thousands of potential combinations and an equal number of structural configurations to be tested (Fig. 4.18).

An *FEM* analysis will be carried out for each configuration, in order to evaluate the nodal displacements. The solutions that perform best, in accordance with the established *fitness* function, will constitute the new population.

A part of the source *genome* is selected and undergoes changes and recombinations in order to obtain a new population that stems from the first one. The process is reiterated until a specific threshold has been reached; this is established by the *fitness* function.

---

[3] *Galapagos* is a genetic solver that is integrated into *Grasshopper*. It was developed by David Rutten. Further information available at http://www.grasshopper3d.com/group/galapagos.

Fig. 4.19 Flow chart of the optimization process

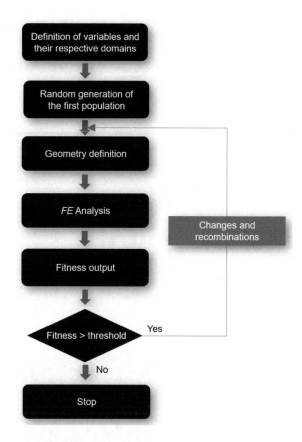

The working diagram of the genetic algorithm can be summarised in the following flow chart (Fig. 4.19).

It is also possible to consider other aspects of construction, such as the minimum and maximum length of the rods or the minimum area of each cell, so that the algorithm automatically rejects the solutions that cannot be produced, even if they are effective in static terms; in the following example, this parameter was not considered, since it refers to a freely designed surface with no dimensional limitations and is solely for illustrative purposes.

The set of variables was defined with two *gene pools*, each of which comprised a number, *n*, of *sliders* equal to the number of cells from the structural *pattern*; the two sets of variables that comprise the *genome* represent the uv coordinates of the centroids of the *Voronoi* cells (Fig. 4.20).

The optimisation process was halted after 26 iterations and made it possible to obtain a defined pattern (Fig. 4.21).

This will not necessarily be the best possible configuration, but the following process will make it possible for the designer to make a more informed decision. Research and a more in-depth study of the structure's behaviour may be carried out at

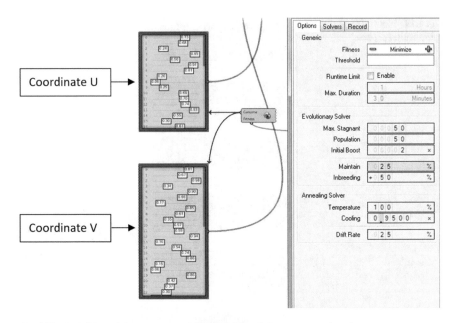

Fig. 4.20 Definition of the genome and setting of the *Galapagos* genetic solver

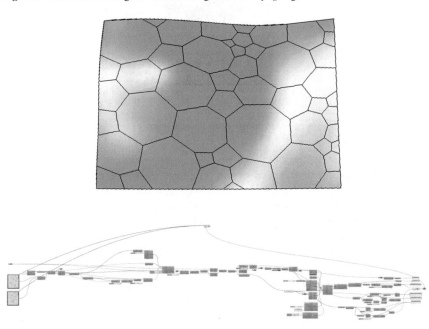

Fig. 4.21 Pattern obtained from the optimisation process after 26 iterations and the complete visual algorithm for the process

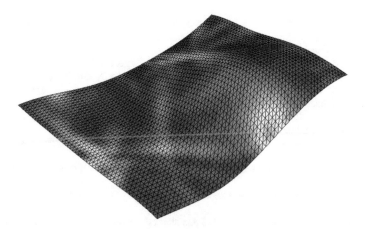

Fig. 4.22 FEM analysis of the meshed surface; areas in blue indicate major Von Mises stresses

a later date, for example, based on the 5 best configurations produced by the genetic solver. As was already mentioned, there are various strategies that make it possible to guide the optimisation process by imposing limitations, such as the minimum and maximum length of the rods or the dimensions of the cells, thus obtaining solutions that are closer to specific construction or aesthetic needs.

It is interesting to note that, in many cases, it is possible to establish a parallel between the structural pattern obtained and the distribution of tension on continuous surfaces (Fig. 4.22).

In terms of output—as well as geometry—all of the useful parameters for evaluating the features of the structure and immediately performing quantitative comparisons have been organised; number of nodes, number of rods, minimum and maximum area of each cell, minimum and maximum length of the rods, sum of the diplacements and stress features.

References

1. Dimcic M (2011) Structural optimization of grid shells. Based on genetic algorithms. PhD Dissertation, University of Stuttgart
2. Mario Bellini Architects (2005) Département des Arts de l'Islam—Musée du Louvre. http://www.bellini.it/architecture/Louvre.html. Accessed 8 Mar 2021
3. Tonelli D (2013) Progettare Involucri di Forma Libera: Una Panoramica sul Tema. http://www2.ing.unipi.it/griff/files/Cm1.1.pdf. Accessed 7 Mar 2021
4. Lee K (2012) Admirant shopping. https://flic.kr/p/cZFE1S. https://flic.kr/p/cZFDVd. Accessed 7 Mar 2021
5. Di Paola F, Pedone P, Inzerillo L, Santagati C (2015) Anamorphic projection: analogical/digital algorithms. J Nexus Netw 17:253. https://doi.org/10.1007/s00004-014-0225-5
6. Wayne T, Piegl L (1997) The NURBS book. Springer, Berlin

7. Pottmann H, Eigensatz M, Deuss M, Schiftner A, Kilian M, Mitra NJ, Pauly M (2010) Case studies in cost-optimized paneling of architectural freeform surfaces. In: Advances in architectural geometry. Springer-Verlag, Vienna

8. Popov EV (2002) Geometric approach to chebyshev net generation along an arbitrary surface represented by NURBS. International Conference Graphicon, Nizhny Novgorod

9. Atsuyuki O, Barry B, Kokichi S, Chiu SN (2000) Spatial tessellation: concepts and applications of voronoi diagrams. John Wiley & Sons, New York

10. Qiang D, Vance F, Gunzburger M (1999) Centroidal voronoi tessellations: applications and algorithms. Siam Rev J 41(4):637–676

11. Coral. http://www.messersmith.name/wordpress/wp-content/coral_img_0864.jpg. Accessed 11 Mar 2020

12. Diatom.   http://upload.wikimedia.org/wikipedia/commons/b/b9/Diatomeas-Haeckel.jpg. Accessed 11 Mar 2020

13. Dragonfly wings. http://dakedesign.com/images/dragonfly_drk-u9469.png. Accessed 11 Mar 2020

14. Cracks in an arid landscape. http://i282.photobucket.com/albums/kk273/BruceDSTaylor/Voronoi/drying_mud.png. Accessed 11 Mar 2020

15. Adriaenssens S, Block P, Veenendaal D, Williams C (2014) Shell structures for architecture: form-finding and optimization. Taylor & Francis, London

Chapter 5
Different Possibilities of Experimentation Design

Abstract The connection joint presented in Chap. 3, *Algorithmic Modelling and prototyping of a connection joint for free-form reticular space structures*, can be used to produce any kind of reticular structure. Although the solution was designed for bamboo structures, the connection joint can be equally used for wooden beams or steel rods and with any kind of section, provided that the ends have a suitable plate for assembly. In order to showcase some of the possible uses, three possible applications will be outlined: a cover with a double-layer spatial lattice, a geodesic dome and a *free-form* pavilion. The site chosen for a graphic simulation of a hypothetical installation of these structures, which is temporary in nature, is the space in front of Building 14 of the Department of Architecture of the University of Palermo, which comprises a quadrangular area of approximately 37 × 44 m. A site was selected that would allow for the use of spaces linked to university teaching, with the intention of promoting synergy between the activities carried out inside the building and those carried out outside. The multi-purpose pavilion was designed with the intention of becoming a meeting place for students, professors and external individuals, who would find a setting for conferences and workshops in the pavilion structure, as well as displaying students' works and studying in a communal environment.

5.1 Parametric Experiments in Temporary Architectures

As often occurs with architecture, experimentation and validation of new solutions take place through the production of temporary structures—the exhibition pavilions. These have a relatively useful, albeit short, life compared to other types of structures. Their chief purpose is didactic or technological in nature and, for this reason, the production costs are a parameter that is often neglected. The flexibility of the functional requirements for this type of structure makes them a potential meeting place for theoretical research and production methods.

Once the event that they have been produced for is over, some pavilions are intended for final disposal; others are disassembled and remade in new settings, prolonging their lifespans.

F. Di Paola and A. Mercurio, *Parametric Experiments in Architecture*, UNIPA Springer Series, https://doi.org/10.1007/978-3-030-96276-0_5

The use of bamboo for its particular features of workability, flexibility, sustainability and recyclability, as well as being easy to find, as previously described in early sections (Chap. 2, *Bamboo as building material*), is proposed as a material that is particular well suited to technical and aesthetic experimentation, especially for temporary structures that are used as exhibition pavilions.

The use of this alternative material has intrigued famous designers and architects like Buckminster Fuller, Frei Otto, Renzo Piano, Arata Isozaki and Kengo Juma, who experimented, in their works, with the harmonious interaction between nature, art and architecture.

Some significant examples are the bamboo pavilions produced by Markus Heinsdorff for the *"Germany-Chinese Esplanade-Moving Ahead Together"* event series, which was organised between 2007 and 2010 in 5 Chinese cities (Chongquing, Guangzhou, Shenyang and Wuhan, China), the aim of which was to promote sustainable urban development (Fig. 5.1). All the structures (*Navette, Diamond, Lotus, Central pavilions*), which were produced with different technological solutions (bamboo, steel and textile frames) have been designed to be disassembled and remade elsewhere, with materials that are completely recyclable and easily replaced. The type of bamboo used is *Phylostachys pubescens*, known as "Mao bamboos" or "Moso" in China (Fig. 5.2) [1].

One of the most innovative installations designed by Heinsdorff, in terms of the technological solutions adopted, is the pavilion presented for the Shanghai Expo in 2010, known as the *"German-Chinese House"*. The eight-metre-high building was constructed over two floors and the supporting structures comprised bamboo with transversal sections of up to 23 *cm*. The bamboo elements came from the *Julong* species, which originates in southern China. It houses exhibition spaces, conference areas and recreational zones (Fig. 5.3) [2, 3].

In 2018, the Colombian architects Simón Vélez and Stefana Simic, who specialise in natural architecure, designed a towering pavilion of 1,000 m^3 for the international photography event *Contemplation*. This was set up on the banks of the Rhone, facing the ancient city of Arles (France). The construction, produced by *VINCI Construction France*, comprises bamboo elements from the *Guadua* species, which comes from Colombia. It has a central room of 400 m^3 and an external walkway of 600 m^3 for contemplative movement. It can also be fully disassembled (Fig. 5.4) [4].

5.2 A Double-Layered Space Pattern

A first examples shows the application of the connection joint to produce a spatial reticular structure based on a double-layer configuration, with squares on diagonal squares. The geometric genesis begins with the regular tessellation, which comprises a squared polygon of a semi-cylinder; the ribbed cylindrical surface will be transformed into a prism (Fig. 5.5) [5].

Fig. 5.1   Pavilions designed by Markus Heinsdorff for the *"Germany and China—Moving Ahead Together"* event series (Reproduced from Minke [2])

5.3   A Geodetic Dome

Geodesics are extremely resistant structures in proportion to their weight and comprise non-deformable triangular mesh. The first studies that compared the static behaviour of a spatial reticular structure with the geometric tessellation of a sphere were conducted starting from the early twentieth century. The most famous spatial structures were produced by the American architect Buckminster Fuller, who

Fig. 5.2 Diamond Pavilions designed by Markus Heinsdorff for the *"Germany and China—Moving Ahead Together"* event series (Minke [2])

Fig. 5.3 *"German-Chinese House"*, designed by Markus Heinsdorff, Shanghai, China, 2010 (Reproduced from Vidiella [3])

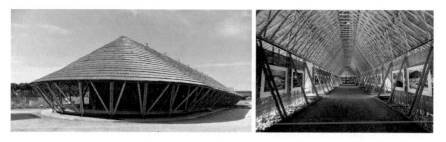

Fig. 5.4 Contemplation Pavilion, designed by Simón Vélez and Stefana Simic, Arles, France, 2018 (Reproduced from Vélez [4])

Fig. 5.5 Front (top) and views (centre and bottom) of the connection joint solution applied to a double-layer reticular configuration

proposed a dome that was about 60 m high, with aluminium rods, for the *Montreal Exhibition* in 1967.

Some recent experiments carried out by *Be Bamboo Design and Construction* propose solutions to modular structural elements of geodesic domes made from bamboo culms, with steel connections that have multiple uses (disaster relief, cultural and entertainment events, tends and recreational habitats). The resistant structure makes it possible to replace the damaged or worn components (Fig. 5.6) [6].

The five regular Polyhedra and 13 Archimedes' Polyhedra can be considered as the only ways to reduce the sphere according to regular polygons with edges of the same length. It is noted, however, that, among the 18 spherical equi-partitions, only the tetrahedron, the octahedron and the icosahedron (which consist of triangular polygons) satisfy the relationship:

$$E = 3V - 6$$

where E represents the number of edges (rods) and V represents the number of vertices (nodes).

There are various ways in which it is possible to generate an approximation of a sphere, transforming the regular polyhedra into polyhedra with triangular faces (Fig. 5.7).

Fig. 5.6 Bamboo dome realized by *Be Bamboo Design and Construction*

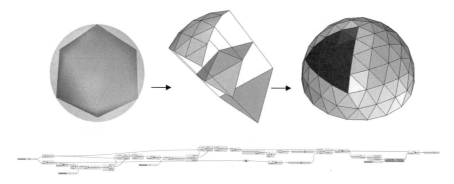

Fig. 5.7 Generation process for a parametric dome with variable levels of subdivision and dimensions

Fig. 5.8 Connection joint solution applied for a geodesic structure

The case study presented here uses the method known as *doubling*: the geometric genesis consists of projecting the central point of the polyhedron's edge from the hundred or spheres that circumscribe said sphere [7].

To design this type of reticular structure with spherical, triangular mesh, it was decided to start with a solid, platonic icosahedron, which, in the field of architecture, best responds to the geometric, technical and aesthetic requirements.

The choice of the type of icosahedral grid is also dictated by the width of the dome in proportion to its height.[1] To obtain our icosahedral grid, two *doubling* operations are applied to the icosahedron (Fig. 5.8).

Although we begin with equilateral triangles, the phase for projecting the points onto the surface of the sphere does not allow for all the rods to be maintained at the same length, even if the variations are minimal.

---

[1] The icosahedron is a platonic solid that comprises 20 equilateral-triangle faces.

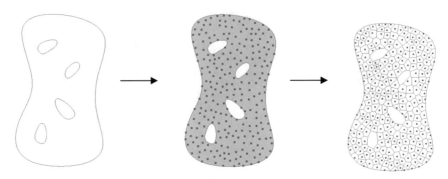

Fig. 5.9 Definition and discretisation of the planar surface by means of *Vonoroi* diagrams

5.4 A *Free-Form* Exposition Pavilion

As previously mentioned, the intention of the design is to create a multi-functional pavilion, one that can accommodate technical-practical learning activities, conferences, workshops and exhibitions in an outdoor environment that serves the University.

The central through line of the design process for the exhibition pavilion was the search for an integrated system that could simultaneously manage the structural forces, the formal aspect and the consequences on the production phases.

In other words, even though it is a *free-form* surface, the designer's choices were never arbitrary; instead, they were guided by static or productive considerations. In this sense, the intention was to bridge the gap between the creative moment and the "engineering" phase of the form. Even during the earliest design phases, *form-finding* makes it possible to check the mechanical behaviour of the structure and, as a result, manage *free-form* surfaces with greater safety and control. *Form-finding* is in no way intended as a substitute for the structural checks that are required for the structure in question, but instead serves as a tool that guides the designer through the creative process.

The modelling of the pavilion starts from the definition of a few geometric parametric inputs, which identify the external edge of the surface and, potentially, the closed planar curves that comprise the internal supports. The latter will be useful for defining a sequence of communicating environments below a single, continuous surface, which have different functions.

The first phase involves randomly generating the coordinates of the points that are to be used as centroids for the cells in the *Voronoi* diagram, in a similar manner to what was outlined for the discretisation of an assigned surface[2] (Fig. 5.9).

We must establish which points of the planar diagram must be maintained at zero—these comprise the external limitations of the structure—and which points will be free to move in the space. In other words, if we want to establish a parallel

---

[2] See Chapter 4, Sects. 4.3 and 4.4.

with the physical models by Gaudi or Frei Otto, we must establish the anchors and the areas of the shell that will be deformed through the forces of gravity. Contrary to what occurs in physical models, the virtual model made it possible to limit these points on a track that comprised the edge curves to which they belong, thus offering greater precision.

It has been noted that this procedure allows the shell to obtain a more effective equilibrium configuration from both a static and formal point of view.

Through the use of attractor points, it was possible to free some of the points that belonged to edge curves, where we want to create openings for design purposes; to prevent these points—which are no longer constrained—from being attracted by forces acting on the shell, they have been limited to moving only along the z axis (Fig. 5.10).

Once the constraints have been established, we must set the *Particle-Spring System* by transforming the geometric entities, lines and points into the corresponding physical entities, elastic springs and particles. The irregular mesh of the Voronoi diagram will thus be able to achieve an equilibrium configuration in which the potential energy of the system is minimised (Fig. 5.11).

However, the process of "relaxing" the shell, which was seen previously in the plan, will be carried out within the space in this case, by introducing an inverse gravitational force that makes it possible to run a physical simulation, similar to those used for real models of hanging shells. In order to iteratively achieve equilibrium for the rod system, the sum of the internal forces F_i (elastic forces) transmitted on each

Fig. 5.10 Definition of the
constraints and lateral
openings of the pavilion

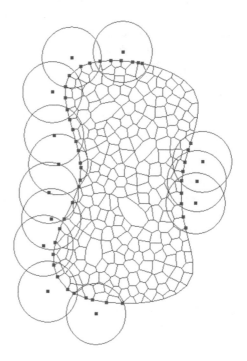

Fig. 5.11 View of the *form-finding* process

node, *i*, must have a result that is equal and opposed to the action of the gravitational force, F_g, that acts on said node. Once equilibrium has been achieved, that is, once the kinetic energy of the system reaches zero, for each node, we must have:

$$\sum \vec{F_i} = \vec{F_g}$$

According to the same diagram that was shown in Chap. 4, an optimisation phase will be carried out at this point. This will occur by means of the *Galapagos* genetic algorithm, which will make it possible to test the static response of the structures, which are generated from hundreds of possible centroid layouts for the *Voronoi* cells.

The algorithmic definition that was developed for the design and optimisation of the free-form exhibition pavilion is outlined below. The block of components linked to the definition of the connection joints, which has already been discussed in detail in previous chapters, has been excluded from this (Fig. 5.12).

By summarising the process, it is possible to identify three key moments in the development of an algorithm: definition of the inputs, definition of the instructions and analysis of the outputs (Fig. 5.13).

In terms of the costs for producing the pavilion, the availability of bamboo must first be evaluated. Even though the costs of importing culms—which typically come from China—are lower, it was decided to refer to Italian producers. For example, for culms that are 60–70 mm in diameter and belong to the *Phyllostachys Viridiglaucescens* species, the price per linear metre is between €2.40 and €5.30,

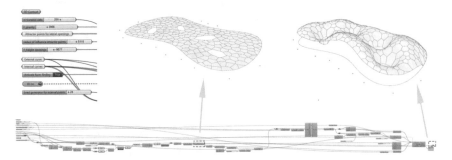

Fig. 5.12 Algorithm for defining the basic surface, the *Voronoi* diagram and for physical *form-finding* simulations

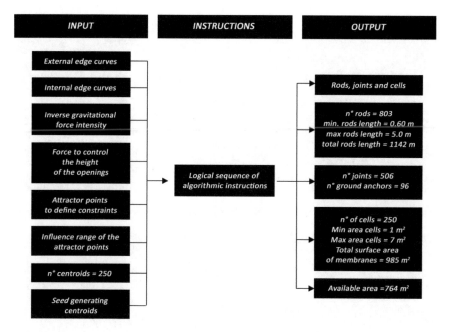

Fig. 5.13   Flow chart: algorithmic definition of the exhibition pavilion

depending on the level of maturity and the treatments applied.[3] Using mature culms that are older than 3 years, mature and which have undergone anti-mold treatments, the cost would be €5.30/metre. Considering the total length of the rods for the pavilion is 1,142 m, the bamboo needed to produce the entire structure would cost just over €6,000, excluding costs for waste and transport. The connection joints would have a decisive influence on the costs of manufacturing; the unit cost for each of these, which are made from CNC-milled steel, was estimated at approximately €220/each, in the case of limited production of a few dozen units[4]; however, larger-scale production would reduce costs (Figs. 5.14, 5.15, 5.16 and 5.17, 5.18).

---

[3] Estimate of costs obtained from Bambuseto, a company based in the province of Lucca. Further information available at http://www.bambuseto.it/.

[4] Estimate obtained from Skorpion Engineering Srl, who are based in the province of Milan. Further information available at http://www.skorpionengineering.com/index.php/it/.

Fig. 5.14 Front view of the exhibition pavilion. Department of Architecture, University of Palermo

Fig. 5.15 View of the exhibition pavilion. Department of Architecture, University of Palermo

Fig. 5.16 View of the space intended for conferences, workshops and learning activities. Department of Architecture, University of Palermo

Fig. 5.17 View of the spaces intended for study. Department of Architecture, University of Palermo

Fig. 5.18 View of the spaces intended for exhibitions. Department of Architecture, University of Palermo

References

1. Heinsdorff M (2014) Mobile spaces: textile buildings. Jovis, Berlin
2. Minke G (2012) Building with bamboo: design and technology of a sustainable architecture. Birkhauser Verlag Ag, Switzerland
3. Vidiella ÀS (2011) Bambù. Logos, Modena
4. Vélez S, Simic S (2018) Contemplation pavilion. https://contemplation.art/en/simon-velezs-pav ilion/. Accessed 5 Apr 2019
5. Chilton J (2002) Atlante delle strutture reticolari. UTET, Torino
6. http://www.bebamboo.eu/. Accessed 5 Apr 2019
7. Migliari R, Baglioni L (2009) I poliedri regolari e semi regolari con un approfondimento sulle cupole geodetiche. In: Geometria descrittiva V2—tecniche e applicazioni. CittàStudi, Novara, p 356–366

Chapter 6
Conclusions and Further Developments and Appendix

6.1 Conclusions and Further Developments

The work carried out primarily intends to propose a new connection joint system for bamboo structures. The existing connection technologies often pre-determine specific junction angles, forcing the designer to follow standardised patterns and configurations. The proposed solution attempts to reconcile the structural requirement with the aesthetic factor, while ensuring the adaptability of the system to completely generic configurations; this makes it possible to create *free-form* surfaces in which each node is necessarily different from the others.

The use of numerical control machines enables the production of a series of connection joints for which standardisation to a single type does not actually lead to a significant reduction in costs: the manufacture of identical or different connection joints is irrelevant in the case of CNC techniques.

However, nothing prevents the production of only 5–10 standardised types: this would allow the connection joints to be used for a large number of possible configurations and to reuse them for new experiments after the original structure has been dismantled. A use of this type would be applicable, for example, to a series of didactic activities aimed at self-construction and learning the use of bamboo as a building material.

Supplementary Information The online version contains supplementary material available at (https://doi.org/10.1007/978-3-030-96276-0_6). The videos can be accessed individually by clicking the DOI link in the accompanying figure caption or by scanning this link with the SN More Media App.

The ease and speed of assembly and disassembly of the structures, even by unskilled labour, as well as the possibility of replacing individual elements, are among the strengths of the proposed system; these characteristics make it possible to create a travelling exhibition pavilion that can be transported and reinstalled with great ease in various sites, for example following the schedule of a series of travelling events and exhibitions.

In order to validate the effectiveness of the proposed solution, further research will have to be aimed at verifying the resistance of both the plate-rod connection system and the connection joint itself. The latter, currently achievable in steel or aluminium by solid milling, could be produced in the near future with rapid prototyping technologies with lower costs and a widespread diffusion of the production network; the enormous growth of the 3D printing market and the continuous development of new materials that are increasingly resistant and biocompatible suggests that in a very short time there will be the possibility of printing one's own structural connection joint in local contexts lacking the availability of industrial plants.

The new additive manufacturing systems meet perfectly with the development of computational parametric design techniques in a synergistic path that makes it possible to diversify production towards individual products that are organic and morphologically similar to nature and optimised for specific needs, of a functional/structural nature, exactly as it happens in the case of the connection joint, definitively overcoming the concept of serialisation and standardisation. In the near future, most industrial processes will have a digital matrix as a generator of governance and production control.

Beyond the connection system conceived, what is most innovative and with the broadest applicability is the algorithmic process that has made it possible to structure and put in relation the project parts by means of nodal diagrams.

The parametric approach and optimisation by means of *genetic algorithms* based on "objective" functions (*fitness functions*), which refer to mechanical, energy, acoustic or any other type of performance, have made it possible to define a design method that is widely valid. The analysis of the performance and efficiency of an artefact, which usually constitute only a final moment of verification, provides, in the algorithmic process, information useful for redefining the project itself: the answer of the solution adopted can be calculated and verified iteratively obtaining infinite design variants as the input parameters change. Through the definition of adequate constraints to the problem and the choice of appropriate "objective" functions, the designer-programmer guides the entire generation and optimisation process.

The methodological processes described in the text made it possible to define both the geometry of the connection joint and the shape and structure of the proposed exhibition pavilion.

6.2 Appendix

Abstract This chapter provides the links to videos of the resolving processes (Grasshopper, Kangaroo, Galapagos).

Grid Optimization on Free-Form Surfaces_*Voronoi* Relaxation on Generic Surface

For the definition of structural patterns for free-form surfaces, a methodology has been developed that can determine a configuration in which the distribution of stresses is uniform, and in which, as an indirect consequence, there is greater uniformity of rod lengths and polygon angles.

Exactly as it happens in nature, to reach a solution in which the potential energy of the system is minimized, it has been used the *Particle-Spring System* to optimize the *Voronoi* diagram. The optimization algorithm developed, maintaining the initial topology of the *Voronoi* diagram, i.e., the relationships between the geometric entities, "relaxes" the mesh while respecting the balance between the internal and external forces. Each node P_0 of the mesh is considered as a hinge, while each of the rods connected to it are axially deformable according to *Hooke*'s law (Fig. 6.1).

Form Optimization of a Structural Joint with Genetic Algorithm. From Concept to Prototype

The design of the shape of the connection joint had as objective, the search for the solution that undergoes less deformation.

To this end, a genetic solver integrated in *Grasshopper*, *Galapagos*, was used, which allows to determine the best solution, with respect to a given objective function, defined "fitness", among a large population of possible candidates. A very diffused

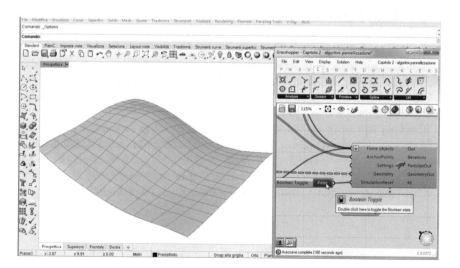

Fig. 6.1 Grid optimization on free-form surfaces (▶ https://doi.org/10.1007/000-7b4)

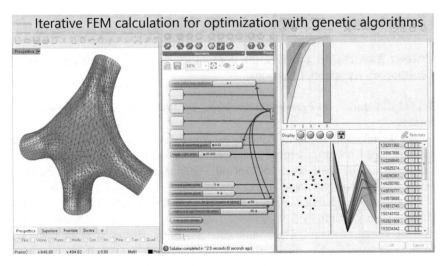

Fig. 6.2 Form optimization of a structural joint (▶ https://doi.org/10.1007/000-7b3)

method to set up the algorithm is that one to minimize, or to maximize, the objective function $f(x)$ going to act on a set of parameters $(x_1, x_2,..., x_n)$ that vary inside an own domain.

The parameters, or the variables on which the algorithm acts, are as mentioned:

- the subdivision of the mesh;
- the smoothing factor;
- length of the arms of the joint and size of the central core (Fig. 6.2).

Form Finding **Technique of the Exhibition Pavilion**

In the case of a priori optimization with *form-finding* techniques, the main objective to be achieved is to transfer loads to the ground by pure compression, minimizing moments. Scale models, in which fabrics or ropes are anchored and subjected to the force of gravity, have long been a useful tool for evaluating the physical and mechanical behavior of a form; this has made it possible to combine formal and structural design in a single process.

The idea was to invert the shape obtained from a rope anchored at the ends, subjected exclusively to traction, to obtain an arch in pure compression. The catenary is the curve according to which a rope is perfectly flexible and subject only to its own weight. The concept of catenary can be extended in space to the case of vaults and shells, obtaining structures in which membrane actions are predominant (Fig. 6.3).

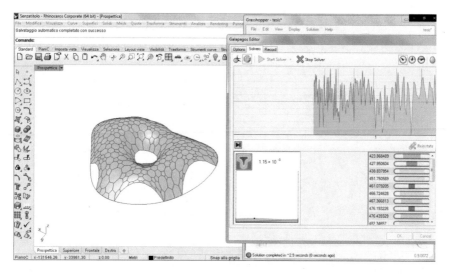

Fig. 6.3 Form-finding of the exhibition pavilion (▶ https://doi.org/10.1007/000-7b5)

Printed in the United States
by Baker & Taylor Publisher Services